Pocket Examiner
in Physiology

Pocket Examiner
in
Physiology

Mary L Forsling BSc, PhD

Senior Lecturer, Department of Physiology,
The Middlesex Hospital Medical School,
University of London.

PITMAN

First published 1981

Catalogue Number 21 1019 81

Pitman Books Limited
39 Parker Street, London WC2B 5PB

Associated Companies
Pitman Publishing Pty Ltd, Melbourne
Pitman Publishing New Zealand Ltd, Wellington

British Library Cataloguing in Publication Data

Forsling, Mary
  Pocket examiner in physiology.
  1. Human physiology—Problems, exercises, etc.
  I. Title
  612'.007   QP40

ISBN: 0 272 79635 2

Set in 9/10 pt VIP Palatino by Herts Typesetting
Services Limited
Printed and bound in Great Britain
at The Pitman Press, Bath

Contents

Preface

An aim of any course in medical sciences is to give the students the basic tools for medical practice, not, as it may sometimes appear to them, to train them to pass examinations. These are, however, hurdles which must be overcome and it is an unwise student who does not prepare himself well. Students generally have very little practice in *viva voce* examinations and in many ways preparation for this form of examination is the most difficult. While purists may object to the aims of this book, nevertheless, it is hoped that it will aid in revising the subject of physiology and be especially useful for oral examinations.

The questions have been selected to represent topics frequently covered in physiology examinations and are in no way intended to represent a syllabus for a physiology course. Emphasis has therefore been placed on the integrated aspects of the subject. A large number of the questions inevitably require factual answers, but many are included to test comprehension. The amount of information required in the answer varies greatly from question to question, but in most instances relatively brief answers have been given. For more detail, the student should consult one or more of the ten standard texts cited or any book recommended for study. These particular textbooks have been selected because they are widely used in medical schools. Physiology is a rapidly developing subject and certain topics are the subject of some controversy, so different textbooks may present different arguments. Areas of dispute have been indicated in the text and the consensus opinion given.

Inevitably some of the answers given are oversimplified; it is not the remit of this book to review all the experimental evidence. For this the student should consult a fuller text and, indeed, is encouraged to do so. Physiology is an experimental subject with changing concepts and it is important that students learn to evaluate the evidence so that they not only comprehend current concepts, but will be able to evaluate those developed 20–30 years hence.

In the course of a *viva*, a student may be asked to clarify a point by drawing a diagram. Some questions therefore ask for this. Diagrams are not included in the text but may be studied in the references cited.

MLF

Some physiological variables and their approximate values in SI units

Parameter	CGS system	SI units
Alveolar air:		
PO_2	14%, 100 mmHg	13 kPa
PCO_2	6%, 40 mmHg	5 kPa
Arterial blood:		
PO_2	95–100 mmHg	13 kPa
PCO_2	40 mmHg	5 kPa
Haemoglobin	14.5 g·100 ml^{-1}	2.2 mmol·l^{-1}
Blood oxygen capacity	20 ml·100 ml^{-1}	8.8 mmol·l^{-1}
Arterial blood:		
O_2	19 ml·100 ml^{-1}	8.5 mmol·l^{-1}
CO_2	48 ml·100 ml^{-1}	22 mmol·l^{-1}
Venous blood:		
O_2	14 ml·100 ml^{-1}	6.0 mmol·l^{-1}
CO_2	52 ml·100 ml^{-1}	24 mmol·l^{-1}
Red cell count	5×10^6·mm^{-3}	5×10^{12}·l^{-1}
White cell count	4000–11 000·mm^{-3}	$4–11 \times 10^9$·l^{-1}
Packed cell volume	45	45
Blood:		
glucose	60–100 mg·100 ml^{-1}	3.5–5.5 mmol·l^{-1}
urea	30 mg·100 ml^{-1}	5.0 mmol·l^{-1}
pH	7.4	7.4
Plasma:		
Na^+	145 mEq·l^{-1}	145 mmol·l^{-1}
Cl^-	105 mEq·l^{-1}	105 mmol·l^{-1}
K^+	5 mEq·l^{-1}	5.0 mmol·l^{-1}
Ca total	10 mg·100 ml^{-1}	2.5 mmol·l^{-1}
Ca^{2+}	2.5 mEq·l^{-1}	1.25 mmol·l^{-1}
proteins	6–8 g·100 ml^{-1}	60–80 g·l^{-1}
Intracellular fluid:		
K^+	150 mEq·l^{-1}	150 mmol·l^{-1}
Mg^{2+}	30 mEq·l^{-1}	15 mmol·l^{-1}
Osmotic pressure of plasma proteins	25 mmHg	4.3–1.6 kPa
Arterial blood pressure	120/80 mmHg	16/11 kPa
Pulmonary artery pressure	25/8 mmHg	3.3/1 kPa
Capillary blood pressure	32 → 12 mmHg	4.3 → 1.6 kPa
Duration of systole	0.3 s	0.3 s
Duration of diastole	0.5 s	0.5 s
Cardiac output	5.0 l·min^{-1}	5.0 l·min^{-1}

Volume of:		
plasma	3.0 l	3.0 l
tissue fluid	12.0 l	12.0 l
intracellular fluid	30.0 l	30.0 l
Glomerular filtration rate	120 ml·min^{-1}	0.12 l·min^{-1}
Urine flow rate	1500 ml·day^{-1}	1.5 l·day^{-1}
Respiratory tidal volume	400 ml	0.4 l
Respiratory dead space	150 ml	0.15 l
Metabolism of:		
1 g carbohydrate	4.1 kcal	17.0 kJ
1 g fat	9.3 kcal	38.0 kJ
1 g protein	4.1 kcal	17.0 kJ
Basic metabolic rate:		
females	37 kcal·m^{-2}·h^{-1}	150 kJ·m^{-2}·h^{-1}
males	40 kcal·m^{-2}·h^{-1}	170 kJ·m^{-2}·h^{-1}

1
Key to references and further reading

A   *Textbook of Physiology and Biochemistry*, Bell, G. H.,
    Emslie-Smith, D. and Peterson, C. R., 9th edn.
    (Churchill Livingstone, Edinburgh, 1980)

B   *Best and Taylor's Physiological Basis of Medical
    Practice*, Brobeck, J. R., 10th edn. (Williams &
    Wilkins, Baltimore, 1979)

C   *Review of Medical Physiology*, Ganong, W. F., 9th
    edn. (Lange Medical Publications, Los Altos,
    California, 1979)

D   *An Introduction to Human Physiology*, Green, J. H.,
    4th (SI) edn. (Oxford University Press, London,
    1976)

E   *Textbook of Medical Physiology*, Guyton, A. C., 5th
    edn. (W. B. Saunders, Philadelphia, 1976)

F   *Wright's Applied Physiology*, Keele, C. A. and Neil,
    E., 12th edn. (Oxford University Press, London,
    1971)

G   *Human Physiology*, Lippold, O. C. J. and Winton,
    F. R., 7th edn. (Churchill Livingstone, Edin-
    burgh, 1979)

H   *Medical Physiology*, Mountcastle, V. B., 14th edn.
    (C. V. Mosby, St Louis, 1980)

I   *Physiology*, Selkurt, E. E., 4th edn. (Little, Brown
    and Company, Boston, 1976)

J   *Human Physiology: The mechanisms of body function*,
    Vander, A. J., Sherman, J. H. and Luciano,
    D. S., 3rd edn. (McGraw-Hill, New York, 1980)

2
Questions

GENERAL PHYSIOLOGY AND BODY FLUIDS

General principles

1 Cells of the human body have some functions in common and some different. Indicate these functions.

2 Explain why diseased cells swell whereas healthy ones do not.

3 State the approximate concentrations of ions inside and outside the cell.

4 Describe the Davson–Danielli model of a membrane.

5 Give the chemical and electrical forces which affect the movement of substances across cell membranes.

6 Give the criteria for facilitated diffusion.

7 Why is active transport an energy-consuming mechanism whereas passive and facilitated diffusion are not?

8 Say why the Donnan effect is important for the distribution of ions across a semi-permeable membrane.

9 Evaluate the importance of Claude Bernard's concept of 'la fixité du milieu intérieur'.

10 What is the difference between negative and positive feedback, and why is the latter uncommon in physiology?

11 Given the context of a system of controls

maintaining a steady internal environment, why do some physiological measures lie within a narrow and others within a wide range?

12 State the Fick principle and its significance.

13 Name two organs through which blood flow can be determined in man, and indicate the markers employed.

14 In what situation other than those listed in answers 12 and 13 may the Fick principle be used?

15 If the oxygen uptake is 225 ml·min^{-1}, the oxygen content of arterial blood 190 ml·l^{-1} and the venous content 140 ml·l^{-1}, what would be the cardiac output?

16 How would the values cited in answer 15 be obtained?

17 Question 16 is concerned with the direct Fick method. What is the principle underlying the indirect Fick method?

18 What other methods are available for determining cardiac output?

Blood

19 Give the approximate value for plasma volume and say how it may be determined.

20 State the concentrations of the main plasma proteins and how they may be separated from plasma.

21 List the chief functions of plasma proteins.

22 Give a reason why oedema may occur in starvation.

23 Describe the formation of lymph and give its functions.

24 Outline the circulation of lymph.

25 In what range does the white cell count normally fall, and how do the various cell types contribute to this number?

26 Bearing in mind the functions of the white cells, in what circumstances and in which direction would you expect to see a change in white cell count?

27 How do the red cell count and the life span of red cells compare with those of white cells?

28 Since red cells have a deep colour and are present in such greater numbers than white cells, how is it possible to carry out a white cell count?

29 What red cell indices may be used clinically?

30 How would red cell indices change with dietary deficiency of iron and chronic blood loss?

31 On examination of a blood sample from a patient, she was found to have a haemoglobin content of 115 $g \cdot l^{-1}$, a red cell count of 3.2×10^{-12} and a mean red cell diameter of 0.2 μm. Are these indices normal and, if not, what would produce such a picture?

32 Outline the life cycle of a red cell.

33 Give details as to the fate of the haem moiety of haemoglobin.

34 In what circumstances is the formation of red cells enhanced and what is the mediator?

35 On what evidence can we say that the red cell has a life of 120 days?

36 Cite the major deficiency in pernicious anaemia and the conditions in which it may occur.

37 For a blood transfusion service, what are the main physiological principles to bear in mind?

38 Name the principal blood group system and explain how the various blood groups are classified.

39 As with many other things in life, why is it that some people are universal donors and others universal recipients?

40 That Rhesus incompatibility leads to haemolytic disease of the newborn is well known. Can ABO incompatibility also occur?

41 In what way is the complement system similar to the blood clotting system?

42 It sometimes happens that a researcher acquires an allergy to the species of animal they study. Explain the underlying mechanisms.

43 Blood groups may be used to determine who is the father of a disputed child. If the mother was O Rh$^-$ and the child A Rh$^-$, could the father be A Rh$^+$ or O Rh$^+$?

44 Bleeding from an artery is far more dangerous than from a vein. Why is this?

45 List the main ways in which bleeding from a small blood vessel may be controlled physiologically.

46 What are the two basic steps in the formation of a fibrin clot?

47 How is prothrombin converted to thrombin?

48 Clarify the use of the terms 'intrinsic' and 'extrinsic' with regard to blood clotting.

49 List the factors associated with blood clotting, indicating which contribute to the intrinsic and which to the extrinsic path.

50 Is it possible for a haemophiliac to suffer from coronary thrombosis?

51 From a knowledge of the clotting mechanisms, list some anticoagulants which could be used to obtain blood for determination of the constituents of plasma. Could these techniques be used *in vivo*?

52 Describe the process of clot dissolution and say why it is important *in vivo*.

Membrane properties of nerve and muscle

53 In what ways may an action potential be measured?

54 Draw a diagram of an action potential in nerve with suitably labelled voltage and time axes.

55 Explain the features of the action potential in terms of the changes in membrane permeability.

56 What happens to action potentials if the sodium–potassium pump is inhibited?

57 In what respect does nervous tissue obey the 'all-or-none' law?

58 Define the terms 'absolute refractory period' and 'relative refractory period'.

59 You wake up in the morning and find an arm has gone 'numb' in the night. What would account for the tingling sensation you subsequently experience?

60 What is the effect on membrane excitability of hypoxia, of introducing local anaesthetics and of reducing the external Ca^{2+}?

61 Is the velocity of the action potential the same in all types of nerve? If not, how does it vary?

62 A compound action potential has a different form if recorded near to the point of stimulation as opposed to a distance from it. Explain why.

63 List the typical conduction velocities for (a) group I muscle afferents, (b) group III or Aγ cutaneous afferents, (c) for group IV or C fibres.

Synaptic transmission

64 Draw a labelled diagram of a synapse.

65 Outline the sequence of events in synaptic transmission and indicate how this transmission may be blocked.

66 List at least six putative neurotransmitters acting centrally or peripherally.

67 List the three features characteristic of synaptic transmission and explain them in terms of the underlying mechanisms.

68 What is the difference between the electrical sign of an inhibitory and an excitatory postsynaptic potential (IPSP and EPSP), and how does it arise?

69 Using the concept of IPSPs and EPSPs explain how 'decisions' are made at synapses.

70 An IPSP produces *direct* inhibition; what other types of inhibition exist?

71 Define convergence, divergence and subliminal fringe.

72 Outline the structure of a muscle end-plate in striated muscle.

73 Describe the effect of acetylcholine on ion permeability at the end-plate region.

74 What is the role of acetylcholinesterase?

75 Explain why drugs such as suxamethonium (succinylcholine) are used during general anaesthesia.

76 Give the effects of curare at the neuromuscular junction.

77 Summarise the changes which section of the nerve produces at the neuromuscular junction.

78 Name a disorder of neuromuscular transmission and indicate the cause.

Muscle contractility

79 Contrast the action potential of skeletal muscle, ventricular muscle and cardiac pacemaker tissue.

80 How does the action potential in smooth muscle compare with that in other muscles?

81 Compare the propagation of impulses in skeletal muscle, heart muscle and smooth muscle.

82 Describe the levels of fibrillar organisation of skeletal muscle and indicate how smooth muscle and cardiac muscle differ.

83 Draw a diagram of part of a skeletal muscle fibril.

84 Explain in terms of the structure how changes in muscle length occur.

85 How can ATP be resynthesised in the muscle?

86 Indicate the significance of triads—i.e. the transverse tubules and neighbouring cisternae of the cytoplasmic reticulum—in terms of Ca^{2+} uptake and release.

87 In simple terms, what is the effect of Ca^{2+} upon actomyosin?

88 Define isometric and isotonic contraction.

89 Explain how a twitch and a tetanus are produced.

90 State the approximate 'time to peak' for the 'twitch contraction' in a fast twitch muscle and a slow twitch muscle.

91 Give the type of curve for a plot of current strength against duration required to depolarise a nerve or muscle membrane to threshold.

92 What is the underlying cause of muscle fatigue?

93 Describe the length tension relationships for resting and contracting muscle during isometric contractions.

94 In terms of the sliding filament, explain why tension is low with very long and very short muscle lengths.

95 How does the force exerted and the work done by a muscle during contraction vary with the speed of contraction?

96 What characteristics do series and parallel elastic components lend to muscle contraction?

97 How does the motor unit allow gradation of muscle contraction *in vivo*?

98 Name the main transmitters affecting smooth muscle and describe their effects.

SYSTEMS OF THE BODY

The cardiovascular system

Heart

99 Give the evidence for the role of the sinoatrial (SA) node as the normal cardiac pacemaker.

100 Say which other sites are potential pacemakers and list them in the ascending order of the frequency of discharge.

101 How does the electrical activity starting at the SA node spread to the atria and ventricles?

102 List the features of atrioventricular conduction and their physiological significance.

103 What is meant by 'heart block'?

104 How may the electrical activity of the heart be determined experimentally and in the intact individual?

105 What type of recording is obtained with the techniques given in answer 104?

106 Explain what an electrocardiogram (ECG) is.

107 State the origin of the term 'Einthoven triangle'.

108 Describe how an ECG may be obtained using standard limb leads.

109 Augmented or unipolar limb leads are used. What are these?

110 Have any other types of leads been used to measure the electrical activity of the heart?

111 Draw a diagram of the ECG and indicate with which events the waves correspond.

112 Given a normal ECG record, what would be the approximate duration of the P–R interval, the QRS complex and the Q–T and the S–T intervals?

113 List some of the information which a physician may obtain from the ECG.

114 Which are the principal types of arrhythmia, and are any seen in the healthy individual?

115 How is the ECG affected if a ventricular beat originates at some ectopic focus?

116 Evaluate the statement 'the heart acts as two pumps'.

117 Define systole and diastole, and give their approximate duration at rest.

118 Draw a diagram of the simultaneous changes in pressure in the left atrium, ventricle and aorta during a cardiac cycle.

119 Mark the events of the ECG on the diagram of the pressure changes.

120 Describe the isometric or isovolumetric phase of contraction, the rapid ejecting and filling phases of the cardiac cycle and state whether the mitral and aortic valves are open or closed.

121 Comment on the pressures in the right ventricle and pulmonary artery during the cardiac cycle.

122 What is the difference between heart sounds and heart murmurs?

123 Name the determinants of cardiac output.

124 Given a cardiac output of 4.2 l·min^{-1} and a heart rate of 70 beats·min^{-1}, calculate the stroke volume and say if the values were obtained in a man or a woman.

125 Give the values for cardiac output, from rest to near maximum exercise, and of heart rate, stroke volume and systolic ventricular volume.

126 Explain what is meant by the chronotropic and inotropic effects of sympathetic stimulation on the heart.

127 State Starling's law of the heart and how catecholamines affect the phenomenon.

128 Enumerate the determinants of heart rate.

129 List the factors affecting stroke volume.

130 From which substrates is the energy for cardiac contraction obtained, and what limitations may result?

131 Name the major determinant of oxygen consumption by the heart.

132 Into what form(s) is the energy of ventricular contraction converted?

133 How is immediate partial compensation for heart failure (reduced cardiac contractility) achieved?

134 In what way does the kidney contribute to additional compensation?

135 Outline the cost to the body of the compensations listed in answer 133.

Circulation

136 Relate the equation describing flow in tubes to the cardiovascular system.

137 Give typical values of systolic and diastolic pressures in the brachial artery of a healthy adult of (a) 20 years and (b) 70 years, and explain how they may be determined by Riva-Rocci's method.

138 How is the mean arterial blood pressure calculated?

139 Make a sketch showing how the blood pressure varies through the cardiac cycle in the arteries, capillaries and veins.

140 Does the linear velocity of blood vary in different parts of the vascular tree?

141 Explain the resistance to blood flow in vessels in terms of Poiseuille's formula and explain why Poiseuille's law cannot be quantitatively applied to the system. Indicate any other factors determining flow resistance.

142 Say what is the function of the series components of the vascular system, the volume

of blood in each section and, hence, why venoconstriction produces changes in stroke volume whereas vasoconstriction produces changes in peripheral resistance.

143 Describe the functional anatomy of the arteries, capillaries and veins.

144 Define arterial pulse. Draw a curve of the arterial pressure changes and indicate how the elasticity of the wall ('hardening of the arteries') influences the speed of the pulse wave.

145 Draw the pattern of the jugular pulse and say with which events the maxima correspond.

146 Explain what is wrong with someone who has more a waves in the jugular pulse than pulsations in the radial artery.

147 Why does a subject with an incompetent tricuspid valve have a giant c wave in the jugular pulse?

148 Outline the mechanisms whereby material crosses the capillary walls.

149 Do arteriolar dilatation and decreased plasma concentration have the same or opposite effects on net movement of fluid across the capillaries?

150 What functions do the lymphatics provide?

151 State the role of the valves in the venous system?

152 Give the main groups of factors controlling the vascular bed.

153 Summarise the neural control of blood vessels.

154 The regulation of blood pressure depends on changes in which two parameters?

155 Which parts of the central nervous system are involved in cardiovascular control?

156 List at least six humoral agents which affect the peripheral circulation and state whether they produce vasodilatation or vasoconstriction.

157 What is meant by the term 'autoregulation'?

158 Local metabolites are said to cause relaxation of vascular smooth muscle. List at least five such substances.

159 Explain the changes underlying active and reactive hyperaemia and say what is the adaptive value of reactive hyperaemia.

160 Under nervous activation, vascular smooth muscle is maintained in a state of contraction. Does it relax completely on denervation?

161 Say what is meant by an axon reflex.

162 Explain what baroreceptors are, where they are found and what the effective stimulus is.

163 Outline the chain of events which take place when the pressure acting on the baroreceptors drops below normal.

164 Name the other groups of receptors responding to stretch in the cardiovascular system.

165 Baroreceptors are important in the regulation of the cardiovascular system. Do they have any effect on respiration and conversely do the chemoreceptors affect the vasomotor and cardiac centres?

166 Compare the peripheral and central effects of hypoxia and hypercapnia.

167 Which organs have a constant blood flow?

168 Why is it desirable that blood flow in the muscles, skin and viscera be very variable?

169 Give a description of the axon reflex and an example of a response it could explain.

170 Comment on the resistance to pulmonary blood flow.

171 How is the response of the pulmonary blood vessels effective in regulating the ventilation/perfusion ratio in the lung?

172 What is the effect of cardiac contraction on coronary blood flow?

173 State the importance of the oxygen utilisation of the myocardium and the β-adrenergic system in the control of coronary flow.

174 Do the vasomotor nerves play a significant role in the regulation of cerebral blood flow?

175 Why does hyperventilation produce dizziness?

176 Angiotensin plays a role in the regulation of renal blood flow. What is it?

177 By how much does the muscle blood flow increase during physical work and how is this increase achieved?

178 On what factors do the colour and temperature of the skin depend, and what function does the skin circulation subserve?

179 Draw a diagram of the circulation of the fetus and describe the changes that take place immediately after birth.

180 Outline the main cardiovascular changes on assumption of the upright posture.

181 Give some causes of systemic hypertension and the physiological principles of treatment.

182 Name some possible causes of hypotension leading to shock.

183 How much blood can be lost by the normal human adult without significant cardiovascular disturbance?

184 Describe the events leading to, and the result of, reduced stroke volume in haemorrhage.

185 What is the effect of haemorrhage on the baroreceptor firing rate and what other receptors are involved in the responses during haemorrhage?

186 Information reaching the medullary centres brings about appropriate compensatory adjustments. How are these achieved?

187 Which hormones may influence the cardiovascular system in haemorrhage?

188 In what ways may respiratory changes in haemorrhage influence cardiovascular changes?

189 How do changes in intestinal flow contribute to redistribution of blood volume?

190 In the course of the long-term compensatory response to haemorrhage, what happens to sodium balance?

191 Give a possible explanation for the thirst experienced in haemorrhage.

192 State the time scale and mechanisms of response for the replacement of fluid, plasma proteins and red blood cells after haemorrhage.

193 Bearing in mind the changes described in the answers to the above questions, what would be the symptoms and signs of severe blood loss?

194 Which methods of treatment may be used for haemorrhage and how effective are they?

The respiratory system

195 What three parameters, important for homoeostasis, does breathing control?

196 Into which four steps can the processes carried out by the respiratory system be summarised?

197 Define and give normal values for vital capacity, tidal volume, inspiratory and expiratory reserve volumes, residual volume and functional residual capacity.

198 How may the functions listed in question 197 be determined?

199 Give a definition of the terms 'hyperpnoea', 'apnoea' and 'dyspnoea'.

200 Why is it impossible to speak if food goes down the wrong way?

201 How does pulmonary ventilation differ from alveolar ventilation?

202 Distinguish between anatomical and physiological dead space, and say whether they have morphological correlates.

203 If you have an anatomical dead space of 140 ml and are out swimming using a snorkel tube of 1.9 cm diameter and 45 cm long, what would your tidal volume have to be to have an alveolar ventilation of 3 $l \cdot min^{-1}$ at 15 breaths$\cdot min^{-1}$?

204 Does the dead space serve any useful function?

205 Where does the chief site of airways resistance lie?

206 Is regional ventilation influenced by posture and, if so, how?

207 Do the alveolar P_{O_2} and P_{CO_2} fluctuate during respiration and, if not, why not?

208 Given that the percentage composition of dried gases in the expired air is oxygen 16 per cent, carbon dioxide 4 per cent, how many ml O_2 would be taken in per minute and how many ml CO_2 produced by a subject expiring 25 litres in 5 minutes (assume volumes to be for dry air at NTP).

209 Say which are the main muscles active during quiet inspiration and which the accessory muscles are.

210 Which muscles are employed during a quiet expiration?

211 Through the respiratory cycle, how do the intrapleural and intra-alveolar pressures change and how are they related?

212 What intrapleural pressures may be achieved in a Valsalva's manoeuvre and in severe coughing?

213 What factors influence the normal position of rest of the lungs?

214 Define compliance and say what would be the lung compliance over the range of intrapleural pressures -2 to -7 cmH$_2$O when 500 ml was inspired, there being no air flow when the pressure determinations were made.

17

215 In terms of the law of Laplace, what should happen to alveoli of different sizes and why is this result not seen?

216 Give the consequences of a deficiency of surfactant.

217 On what principles does measurement of lung compliance depend?

218 What is the effect on compliance of pulmonary congestion, emphysema and asthma?

219 The plot of tidal volume against intrapleural pressure during quiet breathing takes the form of a hysteresis loop. Explain why.

220 Apart from overcoming airway resistance, what else does respiratory work comprise and in what circumstances is the work of breathing increased?

221 State the clinical value of measuring forced expiratory volume.

222 Explain why the distribution of blood through the lungs is not the same through all regions in the upright subject although the pressure difference across the vascular bed is constant.

223 How may the local control of arterioles be said to be similar to that of bronchioles?

224 How does the local control of pulmonary arterioles compare with that of the systemic arterioles?

225 Explain why the systemic arteries have a lower P_{O_2} than blood which has equilibrated with alveolar air.

226 Define the diffusing capacity of the lungs and say what factors influence it.

227 There is increased oxygen diffusion across the alveolar membrane during work. How is this achieved?

228 How may the carbon dioxide and oxygen content of blood be determined?

229 How are gas tensions generally measured?

230 What is the dissociation curve of a respiratory gas, and how may one be prepared?

231 State the difference between oxygen capacity and oxygen content.

232 Describe the form of the oxygen haemoglobin dissociation curve and state why it takes this form.

233 Of what physiological value is the form of the oxygen haemoglobin dissociation curve?

234 Name the main factors which affect the dissociation curve of oxyhaemoglobin.

235 Compared with the amount of oxygen given up from 100 ml blood at a temperature of 37°C and at a P_{O_2} and P_{CO_2} of 40 mmHg (5.3 kPa), would more or less oxygen be given up at (a) a tissue pH of 7.2, (b) a tissue P_{CO_2} or 45 mmHg (6 kPa) and (c) a tissue temperature of 32°C?

236 In what way does the dissociation curve of myoglobin differ from that of haemoglobin, and what is the significance of this difference?

237 List the forms and the amounts in which carbon dioxide exists in the arterial and venous blood.

238 Red cells are normally associated with oxygen carriage, but how do they contribute to the carriage of carbon dioxide?

239 How does a fall in P_{O_2} compare with an increase in P_{CO_2} as a stimulator of respiration?

240 What is the nature of the chemoreceptors and where are they located?

241 Explain why, in carbon monoxide poisoning, respiration is not stimulated despite the fact that little oxygen is carried in the blood.

242 Besides an effect on the respiratory centre, what other effect results from an increase in chemoreceptor activity?

243 Describe the direct effect of changes in P_{aO_2} on the respiratory centre.

244 Apart from the influence of the chemoreceptors, what other afferent inputs to the respiratory centre are there?

245 Where are the centres which have been postulated to control respiration?

246 Outline the present concept of the organisation of the respiratory centres.

247 The term 'ventilation/perfusion mismatch' is used in clinical respiratory physiology. What is meant by the term and what is an important result?

248 Explain why hyperventilation of one area of the lung in response to a low $\dot{V}/\dot{Q}$ will compensate less efficiently for reduced Pa_{O_2} than elevated Pa_{CO_2}.

249 Distinguish between hypoxia, hypoxaemia and cyanosis, and list the four classic categories of hypoxia.

250 Apart from a ventilation/perfusion mismatch, what are the other causes of arterial hypoxia and how may they be distinguished?

251 Explain the immediate steps which should be taken when a patient has acute respiratory and circulatory failure.

252 Oxygen therapy can be very effective in relieving hypoxaemia, but what are the dangers?

The kidney

253 What do you consider are the main functions of the kidney?

254 How does the amount of water ingested as food and drink compare with the urinary output?

255 Draw a diagram of a nephron and indicate the main function of each segment.

256 Nephrons may be differentiated into cortical and juxtamedullary nephrons. What is the main difference between these two types?

257 Indicate on the diagram of the nephron the nature of the renal circulation.

258 Compare the composition of glomerular filtrate and of urine with that of plasma.

259 Define glomerular filtration rate (GFR) and give an average value for man.

260 Write down an equation for the net filtration force and calculate its value, given that the normal intratubular pressure is about 10 mmHg (1.3 kPa) and the oncotic pressure 30 mmHg (4 kPa).

261 Give the effects on GFR of reduced plasma protein concentration as seen in liver disease, circulatory insufficiency and a blockage of urine flow in the ureter.

262 Compare the effects on GFR of constriction of the afferent and efferent arterioles.

263 Clarify the concept of renal clearance and deduce an equation which would allow its calculation clinically.

264 Inulin clearance is used to determine which parameter of kidney function?

265 List the requirements for a substance to be used in the determination of GFR.

266 If the plasma urea concentration is 0.3 $g{\cdot}l^{-1}$, the concentration in urine is 15 $g{\cdot}l^{-1}$ and the urine flow is 1.0 $ml{\cdot}min^{-1}$, what is the plasma clearance of urea? Urea may be used to determine GFR as no injection is required, but which substance is now generally used clinically?

267 Clearance techniques have played a central role in the study of renal function. What other techniques have been important recently?

268 Distinguish between excretion, secretion and reabsorption, and give an example of a substance which is handled by the kidney in each of the second two ways.

269 How may the concept of clearance be applied to the determination of renal blood flow? What other principle may be applied to its determination?

270 Briefly explain autoregulation of renal blood flow.

271 Cite three lines of evidence which have been taken to indicate that a number of substances are actively transported in the kidney.

272 The active reabsorption of sodium affects the movement of other substances. Which are these?

273 Say what is meant by the tubular maximum for glucose, and indicate how you could use a value for it to predict the renal threshold. Is the actual threshold the same as this predicted threshold?

274 The fluid in Bowman's capsule is isotonic with plasma. Give the tonicity of the fluid in the proximal tubule, distal tubule, collecting ducts and ureter, and hence deduce the site at which urinary concentration occurs.

275 The urine excreted may be hypotonic, as the fluid in the distal tubule, or very concentrated, up to $1.3 \text{ osm} \cdot l^{-1}$. How is this high concentration achieved?

276 Indicate briefly how the concentration gradient is maintained.

277 How are amino acids handled by the kidney and what is the significance of the efficiency of this mechanism?

278 To what does the term 'juxtaglomerular apparatus' refer, and what is its possible function?

279 A number of hormones influence the output of urine through actions on the distal tubule and/or collecting duct. Which hormones are these and what is their effect?

280 Describe briefly the mechanics for reabsorption or secretion of hydrogen and bicarbonate.

281 State the various types of diuresis, and indicate which type is produced by diuretic drugs.

282 How are weak acids such as salicylates and phenobarbitone handled by the kidney?

283 In what ways may renal function be affected in pregnancy?

284 Almost any condition which seriously interferes with kidney function can cause acute renal failure. Explain what acute renal failure is and name two of the commonest causes.

285 What are the early features of chronic renal failure?

286 Outline the principles of haemodialysis.

287 Indicate how renal disease may result in hypertension.

288 If the majority of nephrons cease to function, is dialysis always necessary, and, if not, how may the condition be managed?

289 Bladder emptying (micturition) involves the co-ordination of the voluntary and the autonomic nervous system. What three components are involved in bladder emptying and what is their innervation?

290 Of what significance is the law of Laplace with regard to the amount of fluid the bladder will hold before the micturition reflex is brought about?

Digestion

291 List the six essential classes of foodstuff.

292 Give examples of four vitamins and their food sources.

293 Amongst the most important vitamins are A, B, D, riboflavin and ascorbic acid. What is the effect of gross deficiency of each of these?

294 Which inorganic substances are essential in the diet?

295 What is the recommended dietary allowance of calcium and iron, and what is their importance in the diet?

296 Energy released in metabolism is converted to heat, stored or used in external activities. Using

the principle of energy conservation therefore, derive an equation for energy balance after food and in the fasting state.

297   Give a definition of basal metabolic rate and average values for a man and a woman.

298   The number of kilojoules (kJ) (or calories) produced depends on the nature of the foodstuff metabolised; i.e. each class of foodstuff has a given energy of calorific value. Explain how this may be measured in a bomb calorimeter and give approximate values for carbohydrate, fat and protein.

299   Write down an equation for the respiratory quotient (RQ), and give its value when carbohydrate, fat and protein are being metabolised.

300   In the course of a practical class on metabolism during exercise, some students reported an RQ of greater then 1.0. Is this possible or was the result due to experimental error?

301   What is meant by the kilojoule or calorie equivalent of oxygen?

302   Bearing in mind the answer to question 301, say how oxygen consumption may be used to determine metabolic rate.

303   Name a direct method of determining metabolic rate.

304   In a class of male and female students, would it be correct to take a mean value for metabolic rate for the whole class? List the main factors influencing metabolic rate.

305   Give an explanation of the phrase 'specific dynamic action' and say for which category of foodstuff it is greatest.

306   Name a single yet obvious piece of evidence that mechanisms exist for regulating and balancing food intake and metabolic activity.

307   Enumerate the factors which influence the intake of food.

308 Given that the equation for surface area is $0.007184 \times weight^{0.425} \times height^{0.725}$, what is the basal metabolic rate of a subject 162 cm tall and weighing 50 kg, if his rate of oxygen consumption is 300 ml·min⁻¹?

309 Cite the main experimental evidence for believing that there are centres in the midline portion and outer hypothalamus which affect appetitite.

310 The gastrointestinal tract has four main activities. What are they?

311 The gastrointestinal tract has the same general structure throughout its length. Draw a diagram illustrating the main features.

312 Summarise the pathways regulating gastrointestinal activity.

313 Compare the activity of the sympathetic nerves supplying the gastrointestinal tract with that of the parasympathetic nerves.

314 Describe the general structure of the salivary glands and their innervation.

315 Is the tonicity of saliva generally more or less than that of plasma, and how is the osmolality affected by the flow rate?

316 Is the ratio of the concentration in saliva to that in serum for Na, Cl, K and HCO₃ respectively greater or less than one?

317 What are the different effects on salivary secretion of sympathetic and parasympathetic stimulation?

318 Name the major stimuli for salivary secretion.

319 Why does food not enter the lungs during swallowing?

320 How is it possible to swallow a sweet while standing on one's hands?

321 Draw a diagram of the stomach.

322 Into which three phases of digestion has gastric secretion been classically divided?

323 Name the secretions of oxyntic (parietal) and peptic (chief) cells, and say how they are controlled.

324 Is the hydrochloric acid secreted into the stomach very strong and how is it produced?

325 How has the control of gastric secretion been studied in dogs?

326 What is the approximate daily volume of gastric juice, and how may the acid output be measured?

327 Gastrin is a hormone of 17 amino acids. Where is it produced and what factors stimulate its release?

328 How does peptic ulceration occur?

329 To treat a gastric ulcer two-thirds to three-quarters of the stomach may be removed. Bearing in mind the chief functions of the stomach, what will be the principal consequences?

330 List the main differences between intestinal smooth muscle and striated muscle.

331 Compare the segmenting and peristaltic contractions of the gastrointestinal tract.

332 What are 'pangs of hunger' and how are these assuaged by taking in food?

333 Say what is meant by the basic electric rhythm of the stomach.

334 How does emptying of the stomach occur and what factors influence it?

335 No food is allowed prior to administration of general anaesthetic. Why is this?

336 The stomach produces both an enzyme-rich and an aqueous juice. Is the same true of the exocrine pancreas?

337 Powerful proteolytic enzymes derive from the pancreas. Why is the pancreas itself not digested?

338 How is the composition of the pancreatic juice affected by the nature of the stimulus for secretion?

339 Apart from influencing the production of pancreatic juice, what other actions do secretin and cholecystokinin-pancreozymin (CCK-PZ) have?

340 Describe the effect of loss of pancreatic juice and say how it may occur.

341 Give an account of a relatively simple test of pancreatic secretion.

342 Approximately how much bile is secreted per day and what is its composition?

343 Are the terms 'bile pigment' and 'bile salt' synonymous?

344 Distinguish between chloretics and cholagogues and give one example of each.

345 Explain the nature and physiological value of the enterohepatic circulation of bile salts.

346 The digestion and absorption of fats is a complex physicochemical process. Explain why and outline the steps involved.

347 Give the importance of bile salts in the digestion and absorption of fats.

348 In what circumstances is plasma bilirubin elevated and in what ways does this occur?

349 How are gallstones formed and in what circumstances may their presence be asymptomatic?

350 The small intestine is the site of absorption. What are the special features of the intestine which give a large surface area for absorption?

351 The cells of the intestinal mucosa have the fastest turnover times of any in the body. Outline the mechanisms stating the time scale involved.

352 What would you predict to be the effect of cytotoxic drugs or ionising radiation on the absorptive capacity of the small intestine?

353 Say what is meant by net absorption and give the three steps involved in absorption.

354 There are three mechanisms whereby substances can enter the mucosal cells of the intestine. Name them.

355 Give some examples of methods used to study absorption in the small intestine.

356 How much fluid is absorbed during the day by the gastrointestinal tract, and how does absorption in the stomach compare with that in the intestine?

357 What is the tonicity of the contents of the intestine?

358 Describe the differences in the flux of salt and water when the intestine contains solutions of below and above 210 mmol·l^{-1}.

359 What mechanism has evolved for vitamin B$_{12}$ absorption, and why is its existence important?

360 How and where is iron absorbed?

361 Why may magnesium sulphate be used as a purgative?

362 Which are the main enzymes of the small intestine?

363 In what form are carbohydrates absorbed and how is breakdown to this product achieved?

364 Outline the mechanisms involved in the absorption of carbohydrate and indicate the role of sodium in this process.

365 What are the sources of the protein which is available for absorption from the small intestine?

366 Are proteins and peptides absorbed by the intestine?

367 How are the amino acids transported into the mucosal cells?

368 Bearing in mind the physiological functions of

the small intestine, what would be the effect of its resection?

369 Name the two main functions of the large intestine.

370 Draw a diagram showing the parts of the colon.

371 Describe the movements of the colon.

372 It is said that there is a paradox in the relation between pressures in the colon in diarrhoea and in constipation. What is this paradox?

373 Summarise defaecation in terms of a controlled spinal reflex.

The endocrine system

374 If you suspected that a particular organ served an endocrine function, what experiments would you need to perform to prove your hypothesis?

375 The endocrine and the nervous system are the two control systems within the body. Do they function independently?

376 What is the difference between a hormone and a prohormone?

377 Into what groups, according to structure, may hormones be divided?

378 Which glands produce steroid hormones and which peptide hormones?

379 Does one cell generally synthesise one hormone?

380 Which organs, other than those with a primary endocrine function, also secrete hormones?

381 What is the simplest way in which secretion of a hormone may be controlled?

382 Probably the most complex hormonal control system is that for the trophic hormones. Draw a diagram to illustrate the mechanisms involved.

383 Specificity in the nervous system derives largely

from the anatomical arrangement. How is specificity obtained in hormonal control, when the transmitters circulate in the blood stream?

384 Outline the steps in the 'life cycle' of a hormone.

385 In what ways may peptide hormones exert their primary action? Which is the most common mechanism?

386 Name two peptide hormones which do not produce their primary effects through the mediation of cyclic AMP, and, taking one of the hormones as an example, distinguish between primary and secondary effects.

387 Outline the steps leading to the formation of cyclic AMP.

388 Explain how the formation of cyclic AMP leads to the characteristic response of the hormone.

389 Some hormones are said to exert a permissive action. Explain what is meant by this with reference to a particular hormone.

390 If peptide hormones generally exert their effect by binding to the cell membrane, where do steroids act and how is this possible?

391 It has been suggested that thyroxine acts to alter basal metabolic rate by uncoupling oxidative phosphorylation. Say why this is not the case and put forward an alternative explanation.

392 Explain what a bioassay is.

393 Describe the principles of radioimmunoassay, and indicate the advantages and disadvantages over bioassay.

394 Compare the nature and embryological origin of the anterior pituitary with those of the posterior pituitary.

395 Enumerate the hormones synthesised by the anterior pituitary and say which cell types produce them.

396 The hypothalamic agents controlling anterior pituitary function have been referred to as

releasing or inhibiting factors, but now many are called hormones. On what basis is this distinction made?

397 If an animal is hypophysectomised, what happens to the secretion of the anterior pituitary gland if it is replaced next to the severed pituitary stalk or under the capsule of the kidney?

398 What is the structure of anterior pituitary hormones, and where else are hormones of a similar nature produced?

399 The control of secretion of growth hormone and prolactin differs from that of other anterior pituitary hormones in what way?

400 Compare the effects of excess growth hormone in the child with that in the adult and say what effect growth hormone deficiency would have in the fetus.

401 For what reason is it not surprising that acromegalics suffer from diabetes mellitus?

402 Outline the pattern of release of growth hormone and ACTH over a 24-hour period.

403 What similar effects do ACTH and TSH have on their target organs?

404 How does the influence of growth hormone on bone and cartilage differ from its other actions, and what was the original evidence for this mechanism?

405 Apart from his size, what other observations would lead you to believe that Goliath of Gath suffered from oversecretion of growth hormone?

406 Explain the chief tests of pituitary function.

407 Which are the two hormones released from the posterior pituitary, and how do they differ in structure?

408 Where are the neurohypophyseal hormones synthesised, and in what form?

409 Say which is the main physiological stimulus for the release of vasopressin and which receptors are involved.

410   Outline the action of vasopressin on the kidney.

411   Give the nature of the action of vasopressin on
      the cardiovascular system and say when this
      effect may be important.

412   Which are the target organs for oxytocin and
      what is the significance of its action?

413   Describe the effects of over- and undersecretion
      of neurohypophyseal hormones.

414   Iodine is essential for the synthesis of thyroid
      hormones. How concentrated is iodine in the
      gland compared to plasma and how is this
      concentration achieved?

415   Outline the synthesis of thyroxine (T_4) and
      tri-iodothyronine (T_3), indicating the role of
      thyroglobulin.

416   TSH controls output from the thyroid gland. List
      (more fully than in answer 403) the ways in
      which this control is achieved.

417   If given in identical quantities to a hypothyroid
      patient, which would be the more effective in
      relieving symptoms, T_3 or T_4?

418   Thyroid hormones regulate the oxygen con-
      sumption of the majority of cells. How is this
      achieved?

419   In what way do the symptoms of thyroid
      deficiency differ between the newborn and the
      adult?

420   What is meant by the terms 'endemic goitre' and
      'goitrogens'?

421   Outline the changes seen in hypersecretion of
      thyroid hormones (Graves' disease).

422   Give a definition of long-acting thyroid
      stimulator (LATS).

423   How may thyroid function be assessed?

424   Draw a diagram of a cross-section through the
      adrenal gland, indicating the hormones secreted
      by each region.

425 Removal of the adrenal gland is fatal. Is this because of the lack of secretion from the cortex or from the medulla?

426 In what way is secretion from the adrenal medulla controlled?

427 Give the chemical relationships of dopamine, noradrenaline and adrenaline.

428 Do catecholamines play any part in the control of blood glucose?

429 Draw the basic skeleton of the steroid hormones and say what activities are associated with those molecules containing 21 and 19 carbon atoms, respectively, and which grouping is essential for the activity of all groups of steroids.

430 Which is the main steroid, in man, with glucocorticoid activity, and how is it transported in the plasma?

431 ACTH controls the release of cortisol. Does it have any effect on aldosterone release, and which other factors influence the release of this hormone?

432 What is the difference between primary aldosteronism (caused by adrenocortical carcinoma) and secondary aldosteronism?

433 Bearing in mind the actions of adrenocortical hormone, what are the symptoms and signs of Cushing's syndrome (oversecretion of the hormone).

434 Outline a defect underlying changes in the adrenogenital syndrome.

435 Summarise the changes seen in Addison's disease.

436 Give the total plasma ion concentration of calcium and say in what forms it circulates in the blood.

437 Explain the importance of maintaining constant levels of plasma calcium.

438 What proportion of calcium in the body is found in bones and in what form does it occur?

439 Describe the excretion of calcium and phosphate.

440 In which ways does serum calcium affect the secretion of parathyroid hormone and calcitonin?

441 Is either parathyroid hormone or calcitonin vital for life?

442 How does parathyroid hormone help to maintain blood calcium?

443 Enumerate the actions of calcitonin on bone, on calcium and phosphate concentration in plasma, and on the excretion of hydroxyproline by the kidneys.

444 From what compound is vitamin D derived, and which is the active form?

445 What are the consequences of vitamin D deficiency?

446 Name the primary substance regulated by the mechanisms of the postabsorptive state.

447 The endocrine pancreas produces two hormones which affect carbohydrate metabolism. Which are they and in which cells are they produced?

448 Without going into great detail, describe the chemistry of glucagon.

449 Which hormones are released in the fed state and which in the fasting state, and what effect do they have on blood glucose concentrations?

450 Upon what physiological processes does blood glucose depend?

451 Give some important actions of insulin.

452 Lack of insulin in diabetes mellitus results in hyperglycaemia, water loss, ketosis, acidosis, excessive nitrogen excretion and electrolyte imbalance. Explain this, bearing in mind the actions listed in answer 451.

453 Diabetes mellitus may have a juvenile or a maturity onset. This classification is not rigid, but what is the main difference between them?

454 A glucose tolerance test may be used in studying normal and pathological carbohydrate metabolism. How is this test conducted and the results interpreted?

The reproductive system

455 Say how genetic sex is determined biologically and what methods are used to establish it.

456 Draw a diagram showing the anatomical organisation of the male reproductive tract.

457 Outline the cellular events in spermatogenesis.

458 What is the effect of temperature on spermatogenesis, and in what condition may this effect be seen?

459 Much is talked about anabolic steroids, especially in connection with sport. Name one and outline its metabolic actions.

460 What aspects of male characteristics and sexual behaviour depend on testosterone?

461 Outline the composition of an ejaculate of normal fertile semen.

462 Draw a diagram illustrating the anatomy of the female reproductive tract.

463 In what way is the formation of the ovum similar to that of sperm formation, and how does the ovary change during the menstrual cycle?

464 Describe the endometrial changes in the menstrual cycle and indicate how the days of the cycle are numbered.

465 Give a simple description, accompanied by a diagram, of the hormonal regulation of the menstrual cycle.

466 Change in basal body temperature may be used

as an index of ovulation. In which direction is the change and how is it produced?

467 Outline the effect of oestrogen on the external genitalia, vagina and breasts.

468 Give the effects of progesterone on the organs listed in question 467.

469 The contraceptive pill is generally a 'combination pill'. Bearing in mind the control of the menstrual cycle, what are the components likely to be and how might they act?

470 Explain the hormonal changes seen at the menopause.

471 At what stage in the menstrual cycle can fertilisation of the ovum occur, and how soon afterwards does implantation follow?

472 How does the corpus luteum of pregnancy differ from that of the menstrual cycle, and what is the significance of this difference?

473 Which are the main hormonal changes in maternal plasma in pregnancy, and how may knowledge of these be used in the diagnosis of pregnancy and in pregnancy monitoring?

474 The functions of which organs in the fetus does the placenta subserve?

475 Considering all the functions the placenta performs, how does the fetal circulation differ from that of the adult?

476 Outline the hormonal events in the stages of labour and say how this knowledge may be used in the induction of labour.

477 Say how the hormonal regulation of milk secretion differs from that of milk let down.

478 Name the differences in composition between cow's and human milk which need to be taken into account if cow's milk is to substitute for human's in feeding.

479 Care should be taken to keep a baby warm, as control of heat regulation is poorly developed. However, a baby has a special mechanism for keeping warm. What is it?

The nervous system

Sensory input

480 Indicate how the quality, intensity and location of a stimulus are encoded by the nervous system.

481 What is the physiological basis of the phrase 'seeing stars'?

482 Classify the different types of sensory receptor.

483 Give five examples of cutaneous receptors and say to what type of stimulus they respond.

484 Define the generator potential and state whether the all-or-none law is obeyed.

485 In what other ways (from that mentioned in answer 484) does the generator potential differ from an action potential?

486 What is the physiological significance of the difference between rapidly adapting (phasic) and slowly adapting (tonic) receptors?

487 Trace the following pathways:
   (*a*)   Dorsal column tracts.
   (*b*)   Dorsal spinocerebellar tracts.
   (*c*)   Lateral spinothalamic tract.
   (*d*)   Anterior spinothalamic tract.

488 Indicate which sensations are conveyed along the tracts listed in question 487.

489 What are the clinical manifestations of a lesion of sensory pathways at the peripheral level and at the supratentorial level?

490 What is the effect of interruption of the spinothalamic tract, and how may this knowledge be used to benefit certain patients?

491 Define the Brown-Séquard syndrome.

492 Explain the term 'sensory homunculus' and say why the face and hands have the greatest representation in the sensory cortex.

The eye

493 Draw a sagittal section of the eye.

494 How is the aqueous humour formed and reabsorbed, and what is the effect of interference with the latter?

495 What is the function of the lacrimal glands?

496 Outline the pattern of refraction of light in the human eye, indicating where the greatest refraction occurs, and say what happens when the lens is removed for cataract.

497 Explain how accommodation occurs.

498 Describe how hyperopia, myopia and astigmatism may be corrected.

499 Most people aged over 50 need reading glasses. Explain why and say how the power of the lens needed is calculated.

500 Give the effects of light, adrenaline and an anticholinesterase on the pupil.

501 Describe rods and cones, and indicate through which regions light must pass before reaching them.

502 Give the principal theory of colour vision and, using it, describe the possible basis of colour blindness.

503 Name the pigment in rods, its parent molecule and indicate for which wavelength absorption is maximum.

504 Define visual acuity and how it may be tested.

505 A visitor in the countryside, you pick out some posts painted green as a landmark. Returning home on a moonless night you notice that they appear black and their outline is blurred. Why is this?

506 Enumerate the effects on the visual field of lesions at the level of the optic nerve, optic tract and optic chiasma.

The ear

507 Draw a diagram of the ear.

508 What is the threshold of hearing, and how does it vary with the frequency of sound?

509 How are loudness and pitch of sound encoded by the human ear?

510 What is the function of the auditory ossicles and the result of its impairment?

511 How does nerve or sensorineural deafness differ from conduction deafness, and how may the two be distinguished?

512 State the function of the vestibular system.

513 How do the receptors in the vestibular system function?

514 What are the symptoms of dysfunction of the vestibular system?

515 How is the source of a sound located?

Taste and smell

516 Into which four groups can the sense of taste be subdivided, and how are these appreciated?

517 Can the ability to discriminate certain tastes be inherited?

518 Why is the taste of food affected when you have a cold?

519 In what ways do the olfactory and gustatory sensations differ?

Pain

520 Does pain have any physiological value?

521 What two fibre types are involved in the transmission of the sensation of pain and what physiological observations does the presence of two pain pathways explain?

522 Name two conditions in which the sensation of pain is enhanced.

523 Give the difference in the nature of cutaneous and visceral pain.

524 List three theories of pain.

525 Define referred pain. Give an example and describe the underlying theory.

526 List clinical approaches to the relief of pain.

The motor system

527 Enumerate the components of the CNS maintaining motor control.

528 Outline the organisation of muscle fibres in a motor unit.

529 Explain how voluntary contraction of a muscle can result in smoothly applied graded force from a few grams up to several kilograms.

530 Define a reflex action.

531 Draw a diagram of a muscle spindle.

532 What is the conduction velocity in the α and γ efferent fibres?

533 Trace the events as the muscle spindle is activated by action potentials in the γ motor neurone.

534 Comment on the statement 'The γ loop has a positive evolutionary value because it will smooth out muscle movement'.

535 Describe the stretch reflex and give two clinical examples.

536 The muscle spindle fibres are in parallel with the

extrafusal muscle fibres. What is the relationship of the Golgi tendon organ to the latter fibres?

537 Say what is meant by muscle tone and how it is maintained.

538 When a muscle contracts, what happens to the antagonistic muscle?

539 Say what the flexor (or withdrawal) and crossed extensor reflexes are.

540 A healthy plantar reflex is said to be flexor. What happens in disease of the motor pathways?

541 List the characteristics of a polysynaptic reflex.

542 Define reaction time and indicate its length for a stretch reflex such as the knee jerk.

543 Outline the features of damage to the final common pathway.

544 Outline the changes after an acute lesion of the spinal cord.

545 Describe the recovery from spinal shock.

546 Give an alternative name for the corticospinal tract and say what its function is.

547 Using answer 546, deduce the effect of damage to the upper motor neurone (neurone I) and say how it differs from that of the lower motor neurone (neurone II).

548 Distinguish between ataxia and apraxia.

549 Why are the degree and distribution of muscle tone important?

550 What are the sources of afferent inputs for reflexes involved in the maintenance of posture and balance in man?

551 Where are the control circuits located for the co-ordination of postural responses?

552 Patients suffering with Parkinson's disease have lowered striatal dopamine levels and lesions of the substantia nigra. On the basis of the answer

to question 551, what would you expect to be the clinical features of the condition?

553 What are the necessary components of locomotion?

554 How is it that a spinal dog can show rhythmic stepping?

555 Does electrical stimulation of the cerebellum lead to sensation or to movement?

556 Draw a diagram showing how interactions of the cerebellum with the two motor systems (pyramidal and extrapyramidal) influence the muscles.

557 The flocculomotor lobe has connections with the vestibular nuclei. What disturbances would you expect to arise from lesions in this region?

558 What is decerebrate rigidity, and how is it affected by cooling of the cerebellum or removal of the anterior lobe of the cerebellum?

The autonomic nervous system

559 Give an alternative name for the autonomic nervous system and explain the classification of the subdivisions.

560 What tissues does the autonomic nervous system supply?

561 In what ways does control of the visceral system differ from that of the motor system?

562 Contrast the sympathetic outflow from the central nervous system with the parasympathetic outflow.

563 How does the sympathetic nerve supply to organs compare with that of the parasympathetic?

564 List the main organs supplied by the parasympathetic fibres in the cranial nerves, and indicate the effects produced by stimulation.

565 In general terms, what is the effect of sympathetic stimulation?

566 What types of lesions produce disorders of function of the urinary bladder?

567 Explain how drugs may be used to distinguish between preganglionic and postganglionic lesions of the pathways supplying the pupil muscle fibres of the eye.

568 The ganglion-blocking drugs produce postural hypotension. Why is this?

569 The 'deadly nightshade' mushroom contains belladonna, an atropine-like drug. What would be the symptoms of someone eating this fungus?

570 What types of visceral receptor are found through the body?

571 By what routes can visceral afferent information reach the central nervous system?

572 Draw a diagram showing a typical sympathetic reflex arc.

573 Where are the centres located which control visceral function?

The reticular activating system and electrical activity of the brain

574 Define the activity of the ascending reticular formation and give its anatomy.

575 Name the principal afferent and efferent connections of the reticular activating system.

576 Point out the result of destruction of the reticular system.

577 Comment on the underlying basis of the EEG. How may it be recorded, and what are the four basic activities?

578 What are evoked potentials?

579 Outline the clinical value of the EEG.

580 Into what two stages may sleep be divided, and how may these be distinguished using an EEG?

581 Briefly give the anatomical components of the limbic system.

582 Outline the function of the limbic system.

583 Cite two effects of stimulating the limbic area electrically.

584 In view of the answer to question 582, what would you expect to be the clinical manifestations of damage to the limbic system?

585 Learning and memory are hard to define precisely, but give a general definition.

586 Conditioned reflexes represent a type of learning. Give an example of unconditioned and conditioned reflexes.

587 In addition to cortical areas associated with visual and auditory information, there are three association areas. Where do they lie and with what aspects of function are they concerned?

588 Explain what is meant by the term 'dominant hemisphere'.

589 Into what two phases can memory be divided? Indicate what the underlying physiological basis might be.

590 Give a definition of retrograde amnesia and say how it might be induced.

591 What is the effect of damage to the anterior pole of the frontal lobes, and of what particular clinical use is this knowledge?

CO-ORDINATED FUNCTIONS OF THE BODY SYSTEMS

Acid–base balance

592 In what terms is the hydrogen ion concentration of a solution expressed?

593 Calculate the pH of solutions whose hydrogen

ion concentrations are 100 $nmol·l^{-1}$ and 50 $nmol·l^{-1}$, respectively.

594 What is the range of pH compatible with life, and the corresponding hydrogen ion concentration?

595 State the three mechanisms by which the body maintains the pH within relatively narrow limits, and indicate the time scale over which they operate.

596 Give an account of the blood buffers.

597 Why is it that although the pK of the bicarbonate system is 6.8, it is still an efficient buffer in the body?

598 Over what range may urinary pH be altered in an attempt to maintain the pH of plasma constant, and what is the significance of the renal contribution to acid–base balance?

599 By what mechanisms is the urinary pH reduced to 4.5?

600 What mechanisms allow an alkaline urine to be excreted?

601 How is electrical neutrality maintained in the renal tubular cell as hydrogen ions are actively transported from the cell?

602 What is meant by the terms 'acidosis' and 'alkalosis'?

603 In general terms, how may acidosis and alkalosis arise?

604 What will be the effect of drugs which inhibit carbonic anhydrase?

605 Give a simple means of determining whether a patient is suffering from respiratory acidosis or alkalosis.

606 Apart from pH and P_{CO_2}, what other parameters are needed to characterise the nature of acid–base disturbances?

607 Cite some situations in which acid–base balance would be disturbed.

608 Suggest the type of disturbance indicated by the following observations:
(a) a pH of 7.75, P_{CO_2} of 80 mmHg (10.7 kPa) and base excess of +4.0 mmol·l^{-1};
(b) a pH of 7.05, P_{CO_2} of 88 mmHg (11.7 kPa) and base excess of −10.0 mmol·l^{-1} and P_{O_2} of 34 mmHg (4.5 kPa).

Haemorrhage

(See also questions 182 to 194)

609 What is the circulating blood volume, and loss of what volume can be compensated for by the body?

610 Immediately upon a haemorrhage, haemostatic mechanisms come into play. Which reflexes come into play to maintain blood pressure?

611 In patients after haemorrhage, to what signs do the changes listed in answer 610 lead?

612 Summarise the events leading to the partial restoration of blood volume.

613 Fluid is replaced first after haemorrhage. In what order are the other blood constituents replaced, and what is the likely mechanism involved?

614 Explain why saline is of limited value in treating haemorrhage, and say what other treatments have been recommended.

Responses to exercise

615 How does the efficiency of exercising man compare with that of the steam engine and the internal combustion engine?

616 When an individual starts to walk or run, does the oxygen consumption immediately rise at a rate proportional to the rate of movement, and what happens to oxygen consumption when exercise ceases?

617 How is an oxygen debt accumulated?

618 Are the changes in pH, P_{CO_2} and P_{O_2} sufficient to account for the increased ventilation seen during exercise?

619 In strenuous exercise all the muscle capillaries open, as compared with 12–20 per cent at rest. How is this achieved?

620 What general circulatory adjustments are necessary to allow the large increase in blood flow to the exercising muscle?

621 How is the increase in cardiac output during exercise achieved?

Responses to high altitude

622 What are the main environmental variables experienced at altitude?

623 Above 2438 m (8000 ft), pulmonary ventilation is increased. What is the short-term effect, and how does the response change with acclimatisation?

624 What changes in haematological indices would one expect to see during acclimatisation to altitude?

625 How is the work capacity affected at altitude, and what is the difference in the responses of subjects given a short exposure (as in expeditions) and those living permanently at high altitude and then taken to an even higher altitude? Explain the changes seen.

626 Would you expect any cardiovascular changes at altitude? If so, name them.

627 Will there be any alterations in fluid balance at altitude?

628 The beginning of space travel has emphasised the vulnerability of certain systems to gravitational stress. In what ways are the problems of weightlessness similar to those of patients on prolonged bed rest?

Responses to temperature variation

629 Give an indication of what is meant by core temperature; say how it varies and how it is maintained.

630 List the mechanisms for heat loss and heat gain by the body.

631 Where are the temperature receptors and the centre for thermoregulation located, and what changes occur in fever?

632 On what environmental factors does heat loss depend?

633 Give an account of the role of the circulation in thermoregulation.

634 If salt and water balance and circulatory regulation took precedence over thermoregulatory sweating, what would be the result?

635 Bearing in mind the answer to question 634, what are the dangers of exposure to hot or hot and humid climate, and what precautions may be taken?

636 Briefly list the factors involved in heat acclimatisation.

637 What types of responses will be seen in a cold environment?

638 How may hypothermia occur, and when is the condition induced clinically?

Responses to high pressure environments

639 Deep-sea divers and caisson workers are exposed to high atmospheric pressures. Briefly discuss the effects of high alveolar gas pressures on the body.

640 If a diver spends several hours at a deep level and suddenly returns to the surface, what will be the result?

641 How may decompression sickness be avoided?

642 Name a danger of a rapid descent to depth.

3
Answers

GENERAL PHYSIOLOGY AND BODY FLUIDS

General principles

1 Each cell carries out the processes to allow its separate identity to be maintained—such as transport of material across membranes, extraction of energy, protein synthesis, etc. The cell also performs functions typical of its tissue (muscle, nerve, connective or epithelial) or organ system.
[C:1; H:999; J:1]

2 The solute composition of the cell differs from that of the extracellular fluid, a situation requiring an active sodium pump. If this fails, then the internal ionic composition is not maintained and water enters the cell.
[G:7; H:1103; I:31]

3

Intracellular fluid				
	K^+	155 mmol·l⁻¹	PO_4	100 mmol·l⁻¹
	Na^+	12 mmol·l⁻¹	HCO_3^-	10 mmol·l⁻¹
	Mg^{2+}	12.5 mmol·l⁻¹	Cl^-	8 mmol·l⁻¹

Extracellular fluid				
	Na^+	145 mmol·l⁻¹	Cl^-	110 mmol·l⁻¹
	K^+	5 mmol·l⁻¹	HCO_3^-	27 mmol·l⁻¹
	Ca^{2+}	2.5 mmol·l⁻¹		
	Mg^{2+}	2.5 mmol·l⁻¹		

[F:14; G:7; H:999]

4 The membrane is visualised as a double layer of phospholipid molecules with the hydrophilic portions lying at the two surfaces and covered with a layer of protein.
[B:1-5; F:2; H:999; J:100]

5 Molecules diffuse across membranes under the influence of chemical and, for ionised molecules, electrical gradients. Large and small lipid-soluble molecules may pass the lipid regions of

the membrane and small molecules of all classes are able to pass through pores, whereas large polar and non-ionised substances cannot cross the membrane by diffusion.
[B:1-18; C:9; H:999; I:19]

6 Facilitated diffusion occurs down a concentration gradient, but at a rate greater than could be accounted for by simple diffusion. The transport shows saturation, stereospecificity, competition and countertransport and is subject to inhibition.
[B:1-20]

7 Passive diffusion depends on the random movement of molecules, net diffusion always proceeding from a region of high to one of low concentration. Sugars and amino acids cross cell membranes more rapidly than could be accounted for by passive diffusion so it is suggested that the molecule crosses the membrane bound to a carrier—hence facilitated diffusion, still occurring down the diffusion gradient. Active transport occurs against the concentration gradient and hence is energy-dependent.
[B:1-22; C:11; F:9; H:1000]

8 The movement of ions is governed by concentration and electrical gradients. The presence of non-diffusible anions (protein) inside a cell favours the diffusion of anions and hinders that of cations. Gibbs and Donnan showed that the ions distribute themselves so that the concentration ratios are equal.

$$\frac{[K^+_{in}]}{[K^+_{out}]} = \frac{[Cl^-_{in}]}{[Cl^-_{out}]}$$

At equilibrium there is a slight excess of anions inside the cell so that an electrical potential exists across the cell membrane, the inside being negative.
[D:181; H:1000; I:20]

9 The composition of body fluids must be maintained within narrow limits, a fact which is basic to the whole of physiology and which was appreciated by Claude Bernard. Cannon later extended this concept and called the stable conditions produced by co-ordinated physiological responses 'homoeostasis'.
[C:20; G:104; J:4]

10 Some physiological parameters which must be kept within narrow limits, such as the pH of the blood, are controlled factors. Others, such as the pH of the urine, have no set point and indeed urinary pH is altered to help maintain a constant blood pH.
[G:110; I:201]

11 Control systems comprise an input side to detect conditions (a sensor), a controller to initiate the appropriate changes and an effector or output side. In negative feedback an increase in the output results in a decrease in the input so a steady state is maintained; e.g. an increase in plasma osmolality results in increased vasopressin release, leading to water retention and a fall in plasma osmolality. In positive feedback, the greater is the output, the greater the input. As this situation is unstable, it does not occur in physiological systems except where there is a definite end-point; e.g. in the expulsion of a fetus, cervical distension by the fetus leads to increased oxytocin secretion, which produces increased uterine contractions and in turn greater oxytocin release.
[G:105; I:201]

12 The Fick principle is in essence a reconfirmation of the law of conservation of mass. It states that the amount of a substance taken up by an organ per unit time is equal to the concentration gradient of that substance across the organ times the blood flow. Its value lies in the fact that it can be used to determine the blood flow to an organ. The calculation can be performed using a substance which is either given off or taken up by the organ.
[C:436; G:237; I:267]

13 The Fick principle may be used to determine the rates of blood flow through the hepatic, renal, cerebral and coronary circulations. The infused substances are bromsulphthalein for the liver, PAH for the renal blood flow and nitrous oxide for cerebral and coronary blood flow.
[F:143; G:235]

14 The Fick principle may be used to determine pulmonary blood flow—which is of course equal to cardiac output—using oxygen or carbon dioxide as the marker substances.
[A:160; D:83]

15 Cardiac output is equal to the output of the left ventricle and is given by the equation

$$\frac{\text{Cardiac}}{\text{output}} = \frac{\text{Oxygen consumption } (\text{ml}\cdot\text{min}^{-1})}{\text{AV oxygen difference } (\text{ml}\cdot\text{l}^{-1})}$$

using the values given

$$\frac{\text{Cardiac}}{\text{output}} = \frac{225 \text{ ml}\cdot\text{min}^{-1}}{190-140 \text{ ml}\cdot\text{l}^{-1}} = \frac{225}{50} \text{ l}\cdot\text{min}^{-1}$$

$$= 4.5 \text{ l}\cdot\text{min}^{-1}$$

[C:436; D:82]

16 The oxygen uptake can easily be determined using a spirometer. Any arterial sample may be used to determine arterial composition. The venous blood sample is obtained by means of a catheter introduced into a forearm vein and using a fluoroscope guided through the right atrium, right ventricle and into the pulmonary artery.
[F:107; I:267]

17 In the indirect Fick method, use is made of the fact that if the gas in the lungs is allowed to come into equilibrium with that in the blood, then the gas composition will be the same as that of mixed venous blood. Thus the necessity of taking a venous sample is avoided. The gas in the lungs is allowed to equilibrate with venous blood by intermittent rebreathing into a bag.
[F:236; G:236]

18 Cardiac output may also be determined using the Hamilton dye dilution method. This depends on the fact that if a known amount of dye is diluted to give a certain concentration, then the volume in which it is distributed (in this case the cardiac output) can be calculated. The dye is injected intravenously and during its passage through the lungs is fairly evenly distributed through the blood. Blood samples are collected from an artery when the concentration of the dye is seen to rise and then fall and, after recirculation, to rise again. The average concentration of the dye is calculated and, from a plot of time after injection against the log of the dye concentration, the time of the first passage of the dye is obtained. The flow in this time can be calculated by dividing the amount injected by the average concentration and hence the flow

rate per minute or cardiac output can be determined.
[C:437; F:106]

Blood

19 The plasma volume is about 3 litres and can be determined by calculating the volume in which an injection of serum albumin labelled with ^{125}I or a dye, Evans' blue, is distributed.
[A:104; G:118; H:1150]

20 One litre of blood plasma contains 70–80 g protein which largely comprises 50 g albumin and 20 g globulin. There is also some 0.3–0.4 g fibrinogen. Albumin and globulin were originally identified, and may be separated by salting out with ammonium sulphate; now electrophoresis is used.
[A:104; I:241; J:257]

21 Plasma proteins exert an osmotic pressure which is important in fluid balance (see question 260). They also transport some hormones, metals and drugs, and act as buffers. More specifically, plasma proteins can be involved in blood clotting and the immunoglobulins act as antibodies.
[A:102; F:15; H:1127]

22 Apart from the high hydrostatic pressure in dependent parts of the body, there are reduced albumin concentrations which lead to reduced absorption of tissue fluid.
[C:454; H:1158; J:294]

23 Lymph is a tissue fluid which enters the lymphatic vessels, its protein composition being less than that of plasma but varying with the region where it is formed. Thus lymphatics serve to remove surplus tissue fluid and protein from tissue spaces. Lymph nodes, found in the lymphatics, are important in the formation of lymphocytes and act as filters. The intestinal lymphatics serve as a pathway for absorption of fat from the gastrointestinal tract.
[G:123; H:284; I:952]

24 The lymphatics arise as blind-ended vessels. Those from the digestive tract join those from the lower limbs to form the thoracic duct, which,

together with the right lymphatic, feeds into the great veins in the neck. The lymph is moved around the system by contraction of the surrounding muscle and the presence of valves in the lymphatics as in the veins.
[C:453; F:152; H:1092]

25 The white cell count varies from 4000 to 11 000·mm^{-3} or 4–11 × 10^9 cell·l^{-1}. The granulocytes (i.e. those leucocytes with granules in their cytoplasm) form 70 per cent of white cells in the adult. Of the granulocytes, about 95 per cent are neutrophils, 4 per cent eosinophils and 1 per cent basophils. The lymphocytes represent 25 per cent and the monocytes 5 per cent of white cells.
[A:113; J:262]

26 The neutrophils are phagocytes, ingesting foreign particles and bacteria, and will be seen in high concentrations in the region of infected tissue. Eosinophils have antihistamine properties and are increased in allergic conditions. Lymphocytes (more particularly B lymphocytes) play an important part in the production of antibodies and their count is increased in chronic infection.
[C:398; D:17; H:1133]

27 The red cell count is considerably greater than that of the white cells, being 5 × 10^6·mm^{-3} or 5 ×10^{12} cells·l^{-1}. The life span is also longer, being 120 days as opposed to a few days for white cells. The average half-life of a neutrophil is 6 hours.
[F:39, 48; J:107, 113]

28 White cells, like red cells, can be counted in a haemocytometer, but instead of being diluted 1 : 200 they are diluted 1 : 20. The diluting fluid is composed largely of water so the red cells haemolyse, and acetic acid and malachite green are used to fix and stain the cells.
[D:16; G:82]

29 The red cell count, haemoglobin content and haematocrit (packed cell volume) are readily determined. From these may be calculated the mean corpuscular haemoglobin (MCH) which is the haemoglobin content of a single red cell, the mean corpuscular volume (MCV) and the colour index (CI). The mean corpuscular haemoglobin

concentration (MCHC), which takes cell size into account, may also be calculated.
[F:30; G:82]

30   Haemorrhage and dietary deficiency of iron both lead to iron deficiency anaemia which does not affect the production of red cells but which leads to a shortage of haemoglobin. The red cell count is slightly lower, but the colour index (normally 0.9–1.1) is low, as is the MCHC which is below 30 per cent (normal range 32–38 per cent).
[I: 244]

31   Both the haemoglobin content and the red cell count are lower than normal, but the mean red cell diameter is greater. This is characteristic of macrocytic or megloblastic anaemia produced by vitamin B_{12} deficiency.
[C:381; D:10]

32   The formation of red cells (erythropoiesis) occurs in the marrow of bones of the skull and trunk in the adult and in all marrow in children. The red cells develop from reticuloendothelial cells, the sequence being proerythroblast, early, inter-mediate and late normoblast, all of which are nucleated. Then reticulocytes are formed which, as their name implies, contain reticulum, and then the red cell, which has no nucleus. After a life of about 120 days the red cells are broken down in the reticuloendothelial system, the bone marrow, spleen and liver, and the constituents of haemoglobin recycled. The amino acids from the globulin part of the molecule and the rest of the cell are used in general metabolism. Part of the haem portion is converted to ferritin and the rest to bilirubin and biliverdin.
[A:107; H:1128; I:244]

33   Part of the haem portion is stored as ferritin for reuse in erythropoiesis. The remainder forms the pigments bilirubin and biliverdin, which, in combination with protein, are transported to the liver and thence via the bile duct to the duodenum as the bile pigments combined with glucuronic acid (see also answer 343). There is some absorption in the intestine, the rest being excreted in the faeces as stercobilinogen and stercobilin. Some of the bilirubin glucuronide is excreted in the urine as urobilinogen.
[C:407; G:186; H:1132]

34 Erythropoiesis is stimulated by oxygen lack such
    as occurs at high altitude or in congenital heart
    disease. It is also dependent on red cell volume,
    the mediator for these responses being eryth-
    ropoietin.
    [C:407; G:186; H:1132]

35 In order to determine the life span of red blood
    cells they are first labelled in some way and then
    injected and their rate of disappearance
    measured. The injected cells may be identified
    by having a different agglutinogen from that
    present in the recipient, or by being labelled with
    ^{31}P or ^{55}Cr. Alternatively, the haemoglobin may
    be labelled with ^{59}Fe, ^{14}C or $^{15}N_2$.
    [A:107; G:83]

36 Pernicious anaemia is due to vitamin B_{12} or folic
    acid deficiency and is primarily a disease of the
    gastric mucosa as the B_{12} in the diet is not
    absorbed in the absence of gastric intrinsic
    factor. Macrocytic anaemia also results from a
    lack of adequate supplies of B_{12} and may result
    from dietary deficiency of B_{12}, from intestinal
    disease such as sprue or following gastrectomy.
    [D:10; F:43]

37 Apart from taking such precautions as guarding
    against cross-infection and selecting suitable
    donors, it is vital when taking blood for
    transfusion that clotting be prevented.
    Smooth-walled plastic is used to collect blood, as
    glass may promote clotting. Citrate is added to
    the blood to remove calcium and thus prevent
    clotting. Glucose is added to the blood as a
    substrate for erythrocyte metabolism and the
    blood stored at 4°C for up to about 3 weeks.
    When blood is to be given, the donor's cells
    should be matched against the recipient's
    plasma.
    [D:19; J:545]

38 The main blood group system is the ABO
    system, in which the blood group is named after
    the antigen present on the red cell—A (42 per
    cent), B (9 per cent), AB (3 per cent) or none
    (blood group O, 46 per cent). Everyone has
    circulating antibodies which develop a few
    months after birth against the antigens which
    they lack.
    [A:130; F:34]

39 If cells with antigen A are infused into someone with anti-A antibodies, the cells are agglutinated; the same is true if group B cells are given to someone with group B antibodies. Clumping of the recipient's cells is not so significant, as the antibodies are diluted. Group O blood (universal donor) may be transfused into patients with any of the four blood groups because the cells have no antigen. Equally, patients with group AB can receive blood from any of the four groups because there are no antibodies in the plasma.
[D14; I:248]

40 If a Rhesus negative mother is immunised by an earlier pregnancy or by transfusion of Rhesus positive blood, then anti-Rhesus antibodies can cross the placenta and agglutinate the red cells of the fetus because the Rhesus antigens are fully developed at birth. Anti-A or anti-B antibodies may also cross the placenta, but the A and B antigens are immature at birth and are widely distributed throughout the tissues, so the incompatibility is not seen or is clinically mild.
[C:410]

41 The complement and blood clotting systems are examples of cascade systems in which a group of proteins normally circulate in the blood in an inactive state, but which can be activated if the first protein in the series is, thus producing a sequential cascade in which active molecules are formed from inactive precursors. The anti-body–antigen complex is a powerful activator in the first step of the complement sequence, which in turn provides mediators for nearly every process in the inflammatory response. If complement fixation occurs on cells, they are lysed.
[A:115; J:530]

42 The response of the antibody-producing machinery depends on whether the individual has been exposed to the antigen previously. The antibody response to first contact occurs slowly. Subsequent exposure leads to a more immediate and marked response. This is called active immunity, and allows the allergic response to the experimental animal to develop on successive exposures.
[G:469; J:546]

43 Blood group O is a recessive characteristic, so blood group A would have to be inherited from the father. Rhesus negative is similarly a recessive characteristic and so if the mother is Rhesus negative, the father could be Rhesus negative or Rhesus positive.
[D:15; F:37]

44 The pressure in the arterial tree is very high, so the body's haemostatic mechanisms cannot control bleeding and immediate intervention is required.
[D:60; I:249]

45 As soon as a blood vessel is damaged it constricts and the opposed endothelial surfaces tend to stick together. The platelets in the blood clump together at the site of injury to form a plug. Finally, after an interval of several minutes, a fibrin clot is formed.
[C:411; H:1137; I:249]

46 In the formation of fibrin, prothrombin is converted to thrombin, which is a proteolytic enzyme converting fibrinogen to fibrin.
[G:470; J:421]

47 Conversion of prothrombin to thrombin is catalysed by an enzyme, the formation of which is the result of a cascade process in which an inactive enzyme is converted to an active form which acts to catalyse the next step and so on.
[D:18; H:1138]

48 Conversion of prothrombin to thrombin can be brought about through the mediation of an intrinsic pathway including collagen and the platelets or an extrinsic pathway which depends on the release of thromboplastin from damaged walls of blood vessels.
[H:1140; I:251; J:117]

49 There are 12 factors associated with blood clotting, numbered I–XIII, there being no VI.

Factor
I   Fibrinogen
II   Prothrombin
III*   Tissue thromboplastin   (*Terms seldom
IV*   Calcium ions   used)
V   Labile factor

		Extrinsic	Intrinsic
VII	Stable factor	✓	
VIII	Anti-haemophilic globulin		✓
IX	Christmas factor		✓
X	Stuart–Prower factor		
XI	Plasma thromboplastin antecedent		✓
XII	Hageman factor		✓
XIII	Fibrin-stabilising factor		

[A:117; D:19]

50 Strange though it may seem, coronary thrombosis has been observed in haemophiliacs, indicating that hypercoagulability is not a prerequisite of the condition.
[J:325]

51 As calcium is essential for blood clotting, a simple way of preventing the process is to precipitate out the calcium ions using fluoride or oxalate. These agents may not be used *in vivo*. A chelating agent which binds calcium, namely EDTA, may also be used. Heparin is another commonly used anticoagulant which affects the action of thrombin on fibrinogen. Heparin is a naturally occurring substance which may be used clinically in the treatment of intravascular thrombosis.
[D:19; G:471; H:1142]

52 Fibrin is broken down by plasmin (fibrinolysin), which is formed from plasminogen through the action of the Hageman factor together with other substances found in the tissue fluid. Thus the same factors are involved in the formation and the dissolution of a fibrin clot. This, however, means that the spread of clotting is prevented and that the new clot slowly dissolves, thus aiding tissue repair. Anti-clotting mechanisms may also prevent clotting in intact vessels.
[A:118; C:412]

Membrane properties of nerve and muscle

53 An action potential may be recorded on an oscilloscope, using a pair of recording electrodes in contact with the outside of a nerve. Alternatively, an internal electrode may be inserted longitudinally into the giant axon of the

squid or transversely into muscle fibres and other cells.
[B:1-40; E:116; G:24; H:49; I:42]

54 The duration of an action potential in mammalian nerves is less than 1 ms. The diagram should show a resting membrane potential of 60–90 mV. Stimulation produces a local potential which reaches a threshold and the potential then rapidly changes and is reversed, the size of the overshoot being 30–60 mV. The potential then returns to resting values.
[B:1-38; C:25; D:181; E:118]

55 In the passage of an action potential there is initially an increase in the permeability to sodium, which enters the axon and produces the overshoot. This change in sodium permeability is very transient and is followed by an increase in potassium permeability and outward movement of potassium ions.
[A:304; E:117; H:56; J:154]

56 It is the ionic differences between the cell and the surroundings which are provided by the Na^+–K^+ pump which are the basis of the excitable properties of nerve and muscle. However, the resting permeability to sodium is low, so when a large nerve or muscle is poisoned it runs down slowly and action potentials are not affected for a period of time.
[A:311; F:255; I:54]

57 The all-or-none law was first observed by Bowditch in 1872 for contraction of the heart. It states that the response in question is not graded, but is either maximal or does not occur at all.
[B:1-43; F:259; G:25]

58 The absolute refractory period occurs when no stimulus, whatever the strength, can initiate a response, whereas during the relative refractory period a strong stimulus can produce a subnormal spike.
[F:259; H:48; J:156]

59 Cutting off the blood supply leads to anoxia so that the sensation could be due to increased potassium permeability or to increased potassium in the extracellular fluid. This leads to

partial depolarisation and the fibre may fire spontaneously.
[A:313; E:125; G:34]

60 During hypoxia, concentration gradients cannot be maintained and the resting potential decreases, which may be associated with spontaneous firing. Local anaesthetics prevent changes in sodium conductance and block the conduction of action potentials. Calcium acts as a membrane stabiliser and its absence results in a reduced concentration gradient for sodium and potassium. Hypocalcaemia therefore reduces the resting potential and increases excitability.
[A:313; E:127; G:34; J:155]

61 Nerve fibres vary considerably in their conduction velocity. The larger the nerve fibres, the greater the conduction velocity. The conduction velocity is also greater in myelinated fibres.
[A:299; C:32; H:77]

62 If the recording electrode is placed near the point of stimulation, the action potential is seen as a single wave. If the recording is made at some distance, three groups of waves are observed, normally labelled A, B and C, with A being subdivided into α, β, γ and δ. This pattern results from the fact that the nerve comprises fibres with different rates of conduction.
[B:1-41; G:35; I:45]

63 Group I fibres (fibre type Aα) have a conduction velocity of 120 m·s^{-1}, group II fibres (Aβ fibres) have a conduction velocity of 3 m·s^{-1} and group IV or C fibres have a conduction velocity of 1–3 m·s^{-1}.
[B:1-41; C:33; D:183; H:76]

Synaptic transmission

64 A synapse is the junction where synapses are transmitted from one nerve to another. The axon or a portion of the first or presynaptic cell ends on the soma, dendrites or some portion of the second neurone, the postsynaptic cell. There are usually swellings or synaptic knobs at the ends of the fibres, and at the site of transmission the presynaptic fibres are filled with granules containing neurotransmitters.
[A:318; C:53; E:614]

65 The action potential in the presynaptic cell releases a neurotransmitter which in turn produces a local or a synaptic potential in the postsynaptic cell. The result may be excitation or inhibition. Transmission may be blocked if transmitter release is blocked, if transmitter binding to the postsynaptic membrane is prevented (competitive inhibition) or if another depolarising agent binds to the membrane (depolarising block).
[E:614; F:262; J:167]

66 Neurotransmitters include acetylcholine, noradrenaline, dopamine, γ-aminobutyric acid (GABA), glycine, glutamic acid, 5-hydroxytryptamine (5-HT or serotonin) and histamine. Recently it has been suggested that peptides such as the enkephalins, endorphins, vasopressin and somatostatin also have a neurotransmitter role.
[A:323; B:1-57; E:615; H:203; J:173]

67 (a) A number of chemical events occur, so there is a delay. (See also anwer 65.)
(b) The anatomical arrangement and specialisation of cells means that only forward conduction occurs.
(c) Because impulses arrive from many sources at any given neurone, synaptic potentials will summate temporally and spatially.
[A:318; C:53]

68 Postsynaptic potentials may be excitatory (EPSP) or inhibitory (IPSP). In the former case the neurotransmitter increases the conductances of sodium and potassium, which brings the membrane potential closer to threshold (i.e. less negative) and so the cell tends to discharge. At other synapses activation results in changes only in potassium and chloride conductance, so the potential is removed from the threshold (more negative) and the neurone becomes less excitable.
[B:1-51; E:618; G:44; H:188; J:170]

69 Activation of a single synapse does not cause firing of a neurone, as the EPSP is too small (0.5 mV). There are many hundreds of synapses on a given neurone and the different EPSPs and IPSPs are summed. If the net effect is depolarisation, the neurone fires; if hyperpolar-

isation, then the response is inhibition. Each synapse in effect has a 'vote'; either the 'ayes' or the 'nays' have it.
[C:56; J:170]

70 Indirect inhibition can occur as a result of previous synaptic discharge. Presynaptic inhibition is also found in which the release of neurotransmitters is reduced. Inhibition is usually produced as a result of synapses from various pathways converging in one neurone. Some cells may inhibit themselves; e.g. in the spinal cord motor neurones give off recurrent collaterals synapsing with an inhibitory neurone or Renshaw cell.
[B:1-54; C:58; D:182; G:45; H:198]

71 Many presynaptic neurones converge on one neurone, hence convergence, an important result of which is facilitation. Equally, most neurones have branches which end on many postsynaptic neurones, hence divergence. If neurones do not fire in response to a stimulus but have their excitability increased, they are said to lie in the subliminal fringe. If presynaptic fibres share postsynaptic neurones, it is possible for the overall response to be reduced, producing occlusion.
[C:53; E:619; J:166]

72 In a neuromuscular junction the presynaptic area comprises a nerve terminal, a muscle fibre and a postsynaptic junction. The nerve in the region of the synapse loses its myelin sheath and lies in a depression in the muscle fibre, the end-feet containing vesicles of acetylcholine. The sarcolemma in this region comprises complex folds.
[B:1-87; C:61; E:148; F:267; J:442]

73 Acetylcholine leaks from the presynaptic terminals in 'quanta' to produce miniature end-plate potentials. A nerve impulse triggers the almost synchronous release of quanta of acetylcholine, which produces an end-plate potential. The permeability of the postsynaptic membrane to both sodium and potassium ions is increased by acetylcholine, leading to depolarisation of the membrane. When the end-plate potential reaches threshold, an action potential in the muscle is triggered.
[B:1-87; D:183; J:224]

74 Acetylcholinesterase acts to break down acetylcholine at the receptor site, ready for partial uptake into the nerve terminal. In the absence of this enzyme the effect of acetylcholine is prolonged. Accumulation of acetylcholine may result in a depolarising block of the muscle membrane.
[B:1-88; D:184; E:149; F:268; H:155]

75 Suxamethonium (succinylcholine) is a depolarising drug used as a muscle relaxant during general anaesthesia.
[F:269; J:224]

76 Curare produces paralysis by binding competitively with acetylcholine binding sites on the postsynaptic membrane, thus preventing the action of acetylcholine.
[A:327; E:150; F:269]

77 When the motor nerve to a skeletal muscle is cut, Wallerian degeneration occurs distal to the site of injury and retrograde to the nearest collateral. There is synthesis or activation of the receptors, so the muscle becomes exquisitely sensitive to acetylcholine—denervation hypersensitivity.
[C:63; G:41]

78 Such disorders include:
(a) Myasthenia gravis, in which there is weakness which increases after exercise. This is due to partial receptor blockade.
(b) Myasthenic syndrome, in which there is a weakness which can be partially overcome by activity. In this condition there is impaired release of acetylcholine from the nerve terminal.
[A:328; E:150; H:177; I:76]

Muscle contractility

79 (a) In skeletal muscle the action potential lasts 2–4 ms, and in cardiac muscle over 200 ms.
(b) The absolute refractory period in skeletal muscle is 1–3 ms, and the refractory period in cardiac muscle over 200 ms.
(c) Skeletal and ventricular muscle have stable resting potentials, whereas in pacemaker tissue it is unstable. The membrane potential declines after each action potential so that a further one is triggered.

(*d*) The mechanical response in skeletal muscle is considerably longer than the duration of the action potential, being 7.5 ms in 'fast' fibres and as long as 100 ms in 'slow' fibres.

(*e*) The result of (*d*) is that cardiac muscle cannot be tetanised, whereas skeletal muscle can. This is obviously important in terms of muscle function. Skeletal muscle needs to build up a sustained contraction, whereas in the regularly beating heart tetanus would be fatal.

[A:139; E:122; J:220]

80   Smooth muscle has no true resting membrane potential; the potential recorded is high when the tissue is inactive and lower when the tissue is active. Spontaneous electrical activity is seen, some of the pacemaker type. Contraction is seen some time after a spike of electrical activity, about 200 ms after the start. If smooth muscle is stretched, there is an increase in the frequency of spikes of electrical activity.

[I:96; J:247]

81   Cells of skeletal muscle are electrically independent. There are, between the cells of smooth and cardiac muscle, areas known as tight junctions where adjacent cells join producing a point of electrical contact, so that impulses are propagated from one cell to another. If a preparation of ventricular muscle is stimulated, the whole preparation is stimulated before repolarisation begins.

[I:88; J:248]

82   Skeletal muscle is composed of thin fibres 1–50 mm long and 10–100 μm in diameter. The fibres in turn are made up of a large number of fibrils 1–2 μm in diameter and these are composed of actin and myosin filaments arranged in such a way as to produce cross-striations. Cardiac muscle also possesses visible striations. The fibres branch and interdigitate, the points of contact being intercalated discs.

[C:35; J:248]

83   One unit of a repeating pattern of striations is called a sarcomere, and is composed of thick myosin filaments and thin actin filaments which overlap to produce an A band. Where the thin filaments exist alone is the I band, across which

runs the Z line formed by strands which interconnect the thin filaments.
[B:1-63; E:132; G:57; H:84; I:84]

84 According to the sliding filament theory of Huxley, the muscle shortens as a result of the relative movement of the thick and thin filaments past each other, the energy for movement coming from the splitting of ATP by the myosin. The sliding movement is produced by the oscillating movements of cross-links formed when bridges on the myosin filaments briefly join onto the actin filaments.
[A:328; B:1-75; E:133; F:243; H:94]

85 ATP is synthesised from ADP, the energy generally being supplied under both aerobic and anaerobic conditions by the conversion of glucose to carbon dioxide and water. Energy can also come from the breakdown of free fatty acids or from the energy-rich compound found in muscle, phosphocreatine.
[B:1-75; C:42; E:140; G:66; H:105]

86 When the transverse tubules are depolarised in an action potential, calcium is released from the lateral sacs of the sarcoplasmic reticulum around the myofibrils, which initiates the process of contraction. Relaxation follows when calcium ions are taken back into the lateral sacs by an energy-dependent process.
[E:139; I:88; J:222]

87 Calcium ions bind to troponin–tropomyosin of the thin actin filaments, thereby freeing the actin to activate the myosin ATPase which releases the energy to produce movement of the myosin cross-bridge.
[A:336; C:39; H:104]

88 An isotonic contraction occurs if a muscle is allowed to shorten under a steady load. In an isometric contraction the muscle does not shorten, but develops increased tension. Neither type of contraction occurs *in vivo*, voluntary movements being a mixture of the two types of contraction.
[D:179; F:244; H:93]

89 If a muscle is experimentally given a stimulus of sufficient strength to activate the fibres, a rapid

rise in tension is produced followed immediately by a fall—a twitch. If a series of stimuli are given at a sufficiently fast rate then the responses fuse into a smooth contraction—a tetanus.
[B:1–79; C:41; D:187; E:968; F:265; H:91]

90 The time between the onset of contraction and the peak of the tension for fast and slow muscles varies in different species. Examples of fast muscles are those of the jaw, with a contraction time in man of 40 ms. The soleus is a slow muscle, with a twitch time of 120 ms.
[A:334; C:38; E:141]

91 For excitable tissues the current strength required to depolarise a nerve or muscle depends on the duration of stimulus and the current strength. The relation between these parameters is hyperbolic. If the time for which the stimulus is applied is too short, there is no response, however strong the current. If the current falls below a minimum strength then there is no response even if it is applied for an infinite time. The minimum current required to produce a response is the rheobase. The time required for twice this current to stimulate the tissue is the chronaxie.
[B:1-38; D:185; F:259]

92 A drop in the force of muscle contraction after prolonged stimulation is known as muscle fatigue. It is dependent on the supply of ATP. White muscle fatigues more rapidly than red because its glycogen stores are quickly broken down. Fatigue in everyday life is mainly due to failure of the muscle fibres to contract. Electrical stimulation of motor nerves, however, can induce a neuromuscular block before muscle contraction fails.
[A:366; H:73; J:234]

93 At the resting length of the muscle, there is little passive tension. As the muscle is stretched, tension rises slowly and then more rapidly. The active tension during a contraction (which is the total minus the resting tension) rises to a maximum at the resting length and then declines.
[A:245; G:62]

94 The dependence of tension production on the muscle length is related to the amount of interaction and hence the overlap between thick and thin filaments. When there is no overlap, there is no interaction between filaments and no tension develops. At very short lengths, the thin filaments begin to overlap, interfering with the formation of cross-bridges.
[E:137; F:243; J:230]

95 A force velocity curve may be obtained showing that the greater the force, the lower the speed of shortening; i.e. the heavier the load, the less quickly it can be lifted. Work (force × distance) increases as the force opposing shortening increases and then falls away again; i.e. there is no work done when the opposing force is zero and when the muscle does not shorten.
[G:62; H:95; I:99]

96 The parallel elastic components give the elastic properties of muscle when it is stretched passively. The presence of series elastic components means that the changes measured in the muscle follow behind the changes produced by the contractile elements. The tension produced by these elements is first transmitted to the non-contractile elastic components and then to the outside of the muscle fibre.
[H:92; J:229]

97 A single motor neurone and the muscle fibres it innervates form a motor unit. An increased frequency of action potentials in the motor nerve leads to summation of contraction of the muscle cells and greater tension is developed. Numerous motor units need to be activated to bring about movement.
[C:44; J:231]

98 Smooth muscle is affected by acetylcholine release from postganglionic endings of parasympathetic nerves and noradrenaline from sympathetic nerves. In vascular smooth muscle, noradrenaline elevates the threshold potential, resulting in contraction; in intestinal smooth muscle, it lowers the threshold potential, producing relaxation. Acetylcholine produces contraction of intestinal smooth muscle.
[C:49; I:188]

SYSTEMS OF THE BODY

The cardiovascular system

Heart

99   There are several pieces of evidence for the role of the SA node as the cardiac pacemaker:
(*a*) Cooling or crushing the node results in bradycardia or slowing of the heart.
(*b*) Application of drugs to the SA node will alter the heart rate.
(*c*) The SA node is the first part of the heart to show electrical activity.
[B:3-50; E:177; F:117; G:263; H:961; I:312]

100   The AV node and the Purkinje fibres are also sites of spontaneous electrical activity and, hence, potential pacemaker cells. When listed according to the frequency of discharge, the ascending order is the Purkinje fibres, the AV node and SA node.
[A:139; E:181; G:263]

101   The cardiac impulse spreads through the right atrium along ordinary myocardial fibres and to the left atrium via the anterior interatrial myocardial bundle (or Bachmann's bundle). The impulses pass to the AV node via the anterior, middle and posterior internodal pathways. There is a specialised conducting system throughout the ventricles, the upper part being the bundle of His and the lower part a complex network of fibres, the Purkinje fibres.
[B:3-49; C:417; E:181; J:269]

102   There is a delay in the transmission of electrical activity from the atria to the ventricles, which allows time for optimal ventricular filling.
  The relative refractory period of the cells of the mid-portion of the AV node is very long, protecting the ventricles from excessive contraction frequencies which could occur if the atria were depolarised at high frequencies.
  The autonomic nervous system exerts its effects on the AV node.
[A:141; E:179]

103   Heart block is a failure of the transmission of impulses in the conducting system.
[C:423; E:184; G:259]

104 The electrical activity of the heart may be recorded experimentally using intracellular microelectrodes, or two electrodes on the surface of the exposed heart or, in man, by two electrodes on the skin.
[A:150; D:30; F:118; G:265]

105 Intracellular records yield a transmembrane potential; the extracellular records give an action potential which resembles the QRST pattern of the ECG obtained using surface electrodes.
[D:30; E: 191; G:264; J:270]

106 The ECG represents the algebraic sum of the fluctuating action potentials of myocardial fibres during the cardiac cycle. The potentials may be recorded at different loci on the surface of the body. Generally, differences are recorded between pairs of points.
[B:3-62; C:417; J:271]

107 Einthoven devised the original electrocardiographic lead system. The vector sum of the electrical activity of the heart at any time was said to lie at the centre of an equilateral triangle bounded by the shoulders and pubic region.
[B:3-64; C:420; H:1008; I:326]

108 The electrical activity of the heart is generally recorded by measuring the potential difference between pairs of electrodes placed at positions equivalent to the angles of the Einthoven triangle, the arms being used instead of the shoulders and the left leg representing the pubic region. The standard leads are:
I   left arm and right arm
II   right arm and left leg
III left arm and left leg
[D:31; E:194; G:267; H:1012; I:326]

109 The limb electrodes are used as in answer 108, but in this system the potential difference recorded is between the potential in one limb and the mean potential in the other two. The leads are named after the limb with the single lead—aVR (right arm), aVL aVF (left foot).
[B:3-66; D:32; E:196]

110 Monopolar limb chest leads may be used when the differences in potential are recorded between

one of six positions on the chest (V_1–V_6) or the limbs (VR, VL, VF) and an indifferent or reference electrode.
[C:420; E:195; I:327]

111 The diagram should show the P wave, corresponding to depolarisation of the atria, the QRS complex corresponding to ventricular depolarisation and the T wave representing ventricular repolarisation. The magnitude and direction of the deflection depend on the leads used in recording.
[A:153; B:3-64; D:28; H:1010]

112 The P–R interval is approximately 0.12–0.2 s, the QRS complex an average of 0.08 s, the Q–T interval 0.4 s and the S–T interval 0.32 s.
[F:120; G:269]

113 Variations of the ECG give information as to disturbances of cardiac rhythm and conduction, the localisation and extent of ischaemic damage, the anatomical orientation of the heart, the relative size of the chambers and the effects of changes in electrolyte concentrations and of drugs.
[B:3-72; G:269; H:1013; I:331]

114 The principal types of arrythmia are sinus arrhythmia, that resulting from disturbed conduction of impulses and that from ectopic foci; the last may result in extrasystoles, paroxysmal tachycardia and, in the case of multiple ectopic foci, fibrillation. Sinus arrhymia occurs during respiration in children and young adults, the heart speeding up during inspiration and slowing in expiration.
[B:3-78; C:442; D:50; E:213; H:1014]

115 A ventricular beat originating at some ectopic focus will appear as an abnormal QRST complex in the ECG.
[G:269; I:333]

116 The heart effectively comprises two pumping systems: the right heart supplying the low pressure pulmonary circulation and the left heart the high pressure systemic circulation.
[B:3-126; F:93; J:264]

117 Systole describes the contraction of the chambers of the heart, and diastole relaxation. At a rate of 75 beats·min^{-1}, systole lasts 0.27 s and diastole 0.53 s.
[D:25; E:167; J:272]

118 At the beginning of systole, the pressure in the left ventricle is that in the thoracic cavity, i.e. zero. During systole it increases to 120 mmHg (16.0 kPa) and then falls to zero during diastole. The elastic recoil of the aorta means that, while the pressure rises to 120 mmHg (16.0 kPa) in systole, it only falls to 80 mmHg (10.67 kPa). Because the atrial walls are thin, the pressure remains close to that of the thorax (−2 to −8 mmHg (−0.27 to −1.07 kPa). It increases by a few mmHg during atrial systole and again in ventricular systole.
[C:432; F:93; H:987]

119 The P wave occurs at the end of diastole, the QRS complex at the beginning of systole and the T wave at the end of systole.
[B:3-104; C:437; D:28; F:93; H:987]

120 At the end of atrial systole, the mitral valve closes. Then follows a period of isometric or isovolumetric contraction when the pressure in the ventricle increases with no change in volume until the increased pressure results in opening of the aortic valve and rapid ejection of the blood. When the ventricle has emptied, the pressure falls and there is a period of isometric relaxation until the pressure in the ventricle is the same as in the atrium and the mitral valve opens.
[B:3-104; G:251; H:987; I:295]

121 The pattern of pressure changes in the right ventricle is similar to that in the left ventricle. The right ventricle obviously pumps out the same amount of blood as the left, but at a lower pressure (0–25 mmHg; 0–3.3 kPa), while that in the pulmonary artery is 8–25 mmHg (1.1–3.3 kPa).
[A:158; C:433; H:982]

122 Two heart sounds are normally heard through a stethoscope in healthy subjects. The first heart sound is caused by closure of the AV valves and the second by closure of the aortic and pulmonary valves.

Murmurs are abnormal sounds heard in various parts of the vascular system when smooth blood flow is disturbed, as in disease of the heart valves.
[A:159; C:435; E:345; G:254; H:989]

123 Cardiac output may be given as the product of stroke volume and heart rate.
[D:35; G:256]

124 The stroke volume in this case is $4200 \div 70$ ml·min^{-1}, i.e. 60 ml·min^{-1}, a value which would be found in a woman.
[I:353; J:275]

125 Heart rate from rest to exercise can lie in the range 60–200 beats·min^{-1}, stroke volume from 65–135 ml over 200 ml or more in severe exercise. End-diastolic volume lies in the range 150–180 ml and falls about 5 per cent in exercise.
[C:440; F:105]

126 Sympathetic stimulation increases heart rate, a chronotropic effect, and the force of contraction, an inotropic effect.
[A:167; F:106]

127 Starling's law states that ventricular performance is proportional to the initial fibre length of the contracting muscle. The relationship is generally presented in the form of a diagram relating end-diastolic volume (fibre length) to systolic pressure or stroke volume, the curve being called the Frank–Starling curve. Sympathetic stimulation shifts the curve upwards and to the left.
[C:438; J:276]

128 Heart rate is determined by the rate of impulse generation in the SA node, which is innervated by sympathetic fibres which speed and parasympathetic fibres which slow the heart. The major influence on the heart rate is the baroreceptor reflex (see question 163).
[A:167; D:50]

129 Stroke volume is affected by two mechanical factors: the atrial pressure (preload), and the arterial pressure, the pulmonary or systemic resistance to blood flow (afterload). Other factors are those which affect the contractil-

ity—heart rate and sympathetic and parasympathetic innervation.
[G:256; J:276]

130   Glucose and fatty acids are the source of energy for cardiac contraction, metabolism being aerobic. The supply of oxygen may be limiting, in which case the work done should be minimised.
[A:142; I:358]

131   Development of muscle tension (work producing pressure) and hence the afterload or arterial pressure is the major determinant of oxygen consumption. In contrast, myocardial shortening (work due to volume) needs relatively little oxygen, so stroke volume is not so important in this respect. However, increases in cardiac contractility result in increased oxygen consumption, so the level of sympathetic activity is important.
[F:108; G:257; H:997]

132   The energy produced in the ventricle is converted into potential energy through the distension of the aorta. If the heart rate is 70 beats·min^{-1} and the blood pressure is 120 mmHg (which is the same as raising the volume of blood 1.5 m) then the potential energy (which is the weight of blood times the distance moved) is 70×1.5 or 105 g metres per beat. The energy is also converted into kinetic energy if the blood is moving, but this is relatively small.
[C:441; F:109]

133   In heart failure, immediate partial compensation is brought about by a fall in cardiac output. This results in reduced arterial pressure, which in turn results in decreased firing of the baroreceptors and increased vasomotor activity. This then leads to increased heart rate, cardiac contractility, peripheral resistance and venous pressure. Hence the cardiac output returns to normal.
[C:493; E:333; I:411]

134   There is renal conservation of sodium, leading to a positive sodium balance and expansion of the blood and interstitial blood volumes, which effect increases in venous and right atrial pressure and cardiac output.
[B:3-23; C:493; E:334; I:411]

135 Autonomic compensation is achieved at the cost of low cardiac reserve and venous compensation at the cost of oedema (expansion of the extracellular fluid). The latter leads to tissue anoxia because of the increased path for the diffusion of oxygen and paroxysmal nocturnal dyspnoea as the fluid shifts from the extremities when the patient lies down.
[B:3-24; C:493; E:339; I:411]

Circulation

136 The equation for flow in tubes is ΛP = flow × resistance, where ΛP is the pressure difference. As the vascular tree is a continuous series of closed tubes, the equation holds for the entire tree. Flow is effectively the cardiac output, the resistance the total peripheral resistance and the pressure difference the aortic minus the late vena cava pressure. Since the vena cava pressure is 0 mmHg, the ΛP is the mean arterial pressure.
[B:3-4; D:39; E:224; H:1018; G:216]

137 The systolic/diastolic pressure is about 120/80 mmHg (16/10.67 kPa) in a 20-year-old. Values increase with age, so at 70 years a value of 160/90 mmHg (21.3/12 kPa) might be recorded. Riva-Rocci introduced an inflatable cuff for recording blood pressure and Korotkov described the method for finding pressures by listening to the sounds in the artery. The cuff is inflated over the brachial artery to above the systolic pressure and the pressure released. The pressure when the first tapping sound is heard is taken as systolic. Eventually sounds become muffled, which is taken as the diastolic pressure in the UK, and disappear, which is taken as diastolic pressure in the USA.
[A:179; B:3-145, 188; D:54; H:1035]

138 The mean pressure is calculated as diastolic pressure plus one-third (systolic − diastolic pressure).
[D:54; J:283]

139 The mean arterial blood pressure of 100 mmHg (13.3 kPa) falls to about 90 mmHg (12.0 kPa) when the arteries are reached, 32 mmHg (4.3 kPa) at the beginning of the capillaries and 12 mmHg (1.6 kPa) at the veins. The difference

between systolic and diastolic pressure is largely lost at the end of the arterioles.
[B:3-7; D:60; E:238; F:68]

140 The linear velocity of the blood, generally expressed in cm·s^{-1}, represents the displacement time of a particle of blood. As the vessels of the cardiovascular system branch, the cross-sectional area of the system increases and so the blood flow velocity in the small vessels is less than in the arteries and veins.
[B:3-7; C:449; I:362]

141 The resistance to flow through a tube depends on the length of the tube (L), the radius of the tube (R), the flow rate (F) and the viscosity (V). The inter-relationship of these factors is combined in Poiseuille's formula in which the pressure drop for the stratified flow of viscous fluids in rigid tubes is defined as follows

$$\Delta P = \frac{8L}{\pi R^4} \Big| FV.$$

This relationship cannot be applied quantitatively, as the blood vessels are not rigid and whole blood is not viscous. Energy may also be lost through turbulence, the critical velocity for turbulence being predicted using Reynold's number.
[A:181; G:216; H:1018]

142 The arteries are transport vessels and contain 20 per cent of the total blood volume; the arterioles are resistance vessels, the capillaries are exchange vessels and contain 8 per cent of the blood volume, and the veins are capacitance vessels and contain 75 per cent of the total blood volume. Thus vasoconstriction (constriction on the arterial side of the tree) increases resistance and venoconstriction (of the venous system) moves stored blood forwards and hence increases stroke volume.
[E:233; G:213; J:282]

143 Arteries have a diameter of the order 0.4 cm (aorta, 2.5 cm) and the wall, which contains a relatively large amount of elastic tissue, is about 1 mm in thickness (aorta, 2 mm). This allows arteries to serve as a pressure reservoir. The lumen diameter of the capillaries is 6 μm and the

wall, which comprises endothelial cells joined at the edges by intercellular cement, is about 1 μm thick. The total surface area is very extensive, allowing effective exchange. The lumen diameter of the veins is about 0.5 cm and the walls, which contain relatively little smooth muscle, are thin (0.5 mm) and are easily distensible. The intima of the veins is folded at intervals to form valves.
[A:174; G:213; I:276]

144 The arterial pulse wave is a wave of pressure which passes rapidly along the arterial system (4–5 m·s^{-1}) at a rate much greater than the velocity of blood flow (0.5 m·s^{-1}). The pulse wave has a dicrotic notch on the descending phase. As the arteries become more rigid with advancing age, the pulse wave moves more rapidly.
[A:178; B:3-21; C:434; E:240; I:368]

145 The jugular pulse shows three maxima, designated a, c and v—a corresponding to atrial systole, c with the pulse wave in the carotid artery and v with the peak in atrial pressure. The minima on either side of v are called x and y, respectively.
[B:3-196; D:29; E:247; F:94]

146 The a waves in the jugular pulse correspond to atrial systole, and the pulse in the radial artery with ventricular systole. In the case of an atrioventricular block, ventricular and atrial systole are dissociated and the atria, for example, may beat at a rate of 100 beats·min^{-1} while the ventricle beats at 40 beats·min^{-1}.
[A:171; B:3-87; C:434; E:185]

147 The c wave corresponds to the pulse in the carotid. The tricuspid valve prevents back-flow of blood from the ventricles to the atria in systole. If this valve is incompetent, the c wave is considerably augmented.
[D:29; J:265]

148 Nutrients and metabolic end-products diffuse across the capillary walls. Bulk flow or ultrafiltration also occurs in which two opposing forces act on fluid movement (the Starling hypothesis). First, hydrostatic pressure forces

fluid out, the driving force being the difference between the capillary pressure (32 mmHg (4.3 kPa) at the start and 15 mmHg (2 kPa) at the end of the capillaries) and the interstitial fluid pressure, which is very low. Second, the difference in osmotic pressure between plasma and tissue fluid (about 25 mmHg; 3.3 kPa) tends to draw fluid in. At the arterial end there is net movement out and at the venous there is net movement inwards.
[H:1089; I:276; J:291]

149 Both arteriolar vasodilatation and decreased plasma protein concentration, as occurs in liver disease, result in increased filtration. When the arterioles dilate, capillary pressure increases as there is less dissipation of pressure in the arteries. Decreased protein concentrations result in decreased plasma osmotic pressure and reduced absorption of fluid into capillaries.
[H:1091; I:283; J:292]

150 The lymphatics serve to return the excess fluid normally coming from the capillaries and to return protein to the blood, as the capillaries may be slightly permeable to protein. In addition, they have specific transport functions to enable substances such as fat to be absorbed from the gastrointestinal tract (see also question 346) to reach the blood. Lymph nodes found in the larger lymphatic vessels play a critical role in the body's defences.
[E:397; F:153; G:123]

151 Venous return is aided by valves in the veins which prevent back-flow in the limbs. Venous return may also be increased by venous smooth muscle contraction, skeletal muscle pressure and respiration.
[C:444; E:243; I:286]

152 Neuronal, hormonal and chemical control is exerted over the various vascular beds.
[B:3-187; G:229; H:1047; J:305]

153 The principal neuronal influence on the blood vessels is via the sympathetic nervous system, which maintains different degrees of vasoconstriction within vascular beds. There is also in muscle a sympathetic vasodilator system. Vasodilatation produced by the parasympathe-

tic in certain structures is thought to result from bradykinin.
[A:192; B:3-164; G:229; H:1047]

154 Blood pressure is regulated through changes in cardiac output and the diameter of resistance and capacitance vessels. A simple equation sometimes quoted is that blood pressure = cardiac output × peripheral resistance.
[D:39; H:1077; J:304]

155 The primary centre for cardiovascular control is the medulla, where the so-called cardiac and vasomotor centres are found. The hypothalamus exerts an important influence on blood pressure, while the premotor cortex and the spinal cord are also involved.
[F:125; G:234; H:1058]

156 Agents affecting the peripheral circulation are:

Circulating catecholamines:	
noradrenaline	vasoconstriction
adrenaline	vasodilatation or vasoconstriction (depending on α or β receptors)
Vasopressin	vasoconstriction
Angiotensin	vasoconstriction
Histamine	vasodilatation in terminal capillary beds, vasoconstriction in larger arterial branches
Bradykinin	vasodilatation
5-Hydroxytryptamine	vasodilatation in peripheral beds
Prostaglandins	vasodilatation

[C:459; E:275; G:232]

157 The capacity of certain organs to maintain their blood flow relatively constant despite alterations in blood pressure is known as autoregulation. It is thought to involve a neurogenic mechanism.
[B:3-8; D:135; E:253; I:398]

158 Such local metabolites include: (a) reduced P_{O_2}, (b) elevated P_{CO_2}, (c) a fall in pH (e.g. lactic acid), (d) adenosine, (e) phosphate, (f) potassium and (g) increased osmolality.
[B:3-163; C:459; E:262]

159 Increased blood flow to a tissue in response to increased metabolic activity is known as hyperaemia. An example of active hyperaemia is the increase in blood flow to exercising skeletal

muscle. Increased blood flow such as that seen when a tourniquet which had obstructed arterial flow is released, is termed reactive hyperaemia. It is due to a chemical substance (or substances) not as yet identified. The adaptive value of the response is that the blood supply to tissue is, to a degree, automatically maintained in the face of ischaemia such as that occurring after partial occlusion of an artery.
[A:190; B:3-217; E:254; J:286]

160   Complete relaxation of vascular smooth muscle is not seen after denervation. This degree of contraction of vascular smooth muscle could be either of myogenic or humoral origin and contributes to residual vascular resistance.
[A:402; G:231]

161   It is thought that sensory nerves may be connected with neighbouring arterioles as well as the sensory organs, so that stimulation of these nerves leads to reflex vasodilatation—an axon reflex.
[A:192; G:230]

162   Baroreceptors are pressure-sensitive receptors located at the bifurcation of the carotid arteries in the carotid sinus and the arch of the aorta. They respond to stretch and, hence, to the level of arterial pressure and to pulse pressure, thus providing added sensitivity because a change in pulse pressure may occur before mean pressure is affected.
[B:3-173; C:461; E:268; F:127; H:1061]

163   As a result of fall in pressure in the region of the baroreceptors, there is reduced nerve impulse activity in the sinus and aortic (depressor) nerves. The inhibitory drive normally exerted on the vasomotor centre is reduced, so there is increased activity in the sympathetic nerves supplying the blood vessels, with resultant vasoconstriction. The reduced input from the baroreceptors also stimulates the cardiac centre, so the heart rate is also increased and the blood pressure returns to normal.
[D:40; E:269; F:127; H:1077]

164   Nerve endings sensitive to stretch are also located in the low pressure or venous side of the circulation in the atria and large veins. They are

said to act as volume receptors and exert an effect on the vasomotor and cardiac centres.
[B:3-182; E:271; G:228; H:1063; I:386]

165   Although baroreceptors are primarily involved in the regulation of blood pressure, they do exert an effect on the respiratory centre, an increase in blood pressure being inhibitory. The peripheral chemoreceptors, sensitive to PO_2, and the central to changes in PCO_2, exert a stimulatory effect on the vasomotor centre.
[C:460; D:43/96; H:1067]

166   Whereas the peripheral effects of hypoxia and hypercapnia are vasodilatation, the central effect is vasoconstriction.
[C:459; D:42; H:1068]

167   Blood flow to the brain remains remarkably constant. Blood flow to the heart is always maintained.
[F:136; H:1094; I:389; J:309]

168   High blood flow to these regions is required only under certain circumstances such as during exercise, digestion or when heat loss is required. In other circumstances, such as circulatory shock, blood can be diverted from these areas to regions where blood supply must be maintained.
[C:479; G:243; H:1101]

169   According to the concept of an axon reflex, sensory nerve fibres in the skin may have collateral fibres passing to adjacent blood vessels. Impulses, in addition to passing to the spinal cord, may pass directly to these blood vessels, causing vasodilatation. This could account for the vasodilatation seen after application of irritants and mechanical stimuli to the skin.
[C:480; D:44; H:1055]

170   The resistance to flow in the pulmonary circulation is low, so the amount of blood flowing through the system can increase two or three times with very little increase in the pressure gradient across the pulmonary vascular bed.
[A:215; B:3-202; D:60; E:313; H:1112]

171 Hypoxia appears to induce vasoconstriction, which results in shunting of blood from inactive parenchyma to regions where gaseous exchange occurs.
[A:218; C:507; E:314; H:1115]

172 During ventricular systole the outflow in the coronary veins is accelerated and the inflow into the coronary arteries is impeded. Overall ventricular contraction impedes flow through the coronary vessels.
[C:476; E:321; H:1099]

173 Low oxygen tension produces coronary dilatation and there is a close relationship between coronary blood flow and myocardial oxygen consumption. The vasodilator substance(s) involved has/have not been identified.
Coronary arterioles contain β-adrenergic receptors which mediate vasodilatation so that when, for example, blood pressure falls—producing reflex increase in adrenergic discharge—there is an increase in coronary blood flow.
[C:478; E:322]

174 The vasomotor nerves play little part in the changes in resistance in the cerebral vessels. The resistance of all cerebral vessels depends on the intracranial pressure. A small decrease of Pa_{O_2} and rise in Pa_{CO_2} reduce arteriolar resistance. This is another example of a local autoregulation system.
[B:3-210; D:203; E:323; F:144; H:1095]

175 It is suggested that an increased sodium load at the macula densa, resulting from increased glomerular filtration, leads to release of renin and, hence, production of angiotensin which induces local vasodilatation.
[A:291; E:375]

176 Angiotensin II may be formed in the kidney, the process being initiated by release of renin from juxtaglomerular cells. This octapeptide causes some constriction of the afferent arteriole, but marked constriction of the efferent arteriole—an effect which leads to a fall in renal blood flow. This drop in the blood flow to the peritubular capillaries leads to a fall in capillary pressure

and, hence, a marked increase in tubular reabsorption.
[A:250; B:3-277]

177 Muscle blood flow can increase by about 30 times during physical work. Initially the rise is brought about by neural mechanisms and later by local factors such as a drop in PO_2, a rise in PCO_2 or a rise in temperature.
[B:3-124; E:370; G:243; J:314]

178 The colour of the skin depends on the state of the capillary bed, and the temperature on the degree of arteriolar dilatation. The skin circulation acts as a blood reservoir and plays a part in temperature regulation.
[B:3-211; C:479; E:380; G:245; H:1101]

179 Blood from the umbilical vein passes to the fetal heart, largely bypassing the fetal liver and travelling in the ductus venosus. The fetal lungs are bypassed by two routes—through the open foramen ovale between the left and right atrium, and through the ductus arteriosus. At birth there is a fall in pulmonary resistance and a rise in the aortic pressure. The ductus venosus and the foramen ovale close, and flow in the umbilical cord ceases. The ductus arteriosus constricts and closes over the next few days.
[E:1125; F:550; H:1966; I:290]

180 On assuming the upright posture, the hydrostatic pressure in the dependent vessels is increased, causing distension, lowered peripheral resistance and displacement of blood into the dependent extremities. The central blood volume and cardiac volumes are diminished and cardiac output is reduced by 20–30 per cent. Blood pressure is maintained by prompt vasoconstriction and there is also an increase in heart rate.
[B:3-199; E:296; H:1039; I:353; J:313]

181 Sustained systemic hypertension results from an increase in peripheral resistance rather than an increase in cardiac output. This may result from renal disease, endocrine disease or coarction (constriction) of the aorta. It may most commonly be termed idiopathic or essential (i.e. of unknown origin). Most anti-hypertensive

drugs act by reducing the sympathetic tone to the peripheral resistance vessels.
[B:3-236; C:494; E:285; I:409]

182 Hypotension may result from a fast effective loss of circulating blood volume, which could be due to haemorrhage, loss of body fluids (e.g. in cholera, burns) or venodilation. It could also result from reduced cardiac function, as with coronary insufficiency or autonomic imbalance, and finally from peripheral vasodilatation, as in toxic hyperaemia, reactive hyperaemia or allergic hyperaemia.
[E:365; G:477; H:1041; J:320]

183 More than 500 ml blood may be lost without significant disturbance.
[A:104; E:358]

184 Haemorrhage leads to a loss in venous pressure which in turn results in decreased venous return, a fall in atrial pressure, a reduction in end-diastolic volume and hence a fall in stroke volume. The decreased stroke volume results in a fall in cardiac output and arterial blood pressure.
[E:360; G:224; H:1081; J:309]

185 The fall in blood pressure leads to reduced baroreceptor firing rate. Haemorrhage also affects atrial receptors and chemoreceptors.
[C:488; H:1081]

186 The reflexes occurring via the medullary centres include a fall in the parasympathetic discharge to the heart and an increase in the sympathetic discharge, which produce a fall in heart rate and an increase in cardiac contractility and stroke volume. The elevated sympathetic discharge to the veins produces an increase in venous tone and hence in venous pressure and venous return, leading to an increased stroke volume. There is also a rise in the sympathetic discharge to the arterioles, excluding the brain and the heart, causing arteriolar constriction and an increase in peripheral resistance.
[C:460; H:1076; J:305]

187 The vasoactive agents vasopressin, angiotensin II and catecholamines are released in haemorrhage; also glucocorticoids.
[B:3-248; C:488; I:481]

188 The respiration rate is increased, which in turn aids venous return.
[A:163; C:486]

189 There is a decrease in venous pressure and an increase in arteriolar constriction with a decrease in capillary pressure which leads to a transfer of fluid from the interstitium to the vascular compartment.
[C:487, 722]

190 After haemorrhage there is a decline in the glomerular filtration rate in urine volume and sodium excretion. This latter is due to haemodynamic changes in the kidney and to increased aldosterone release in response to increased production of angiotensin and release of ACTH.
[C:490]

191 Decreased extracellular volume has been shown to stimulate thirst. Increased concentrations of circulating angiotensin II also play a role.
[C:489; E:479]

192 The plasma volume is restored first within a period of 12–72 hours. Plasma proteins are replaced largely as a result of synthesis in the liver over 3–4 days. The red blood cells are restored in 4–8 weeks. Reduction of blood flow to the kidney stimulates the release of erythropoietin, a hormone which increases the rate of maturation and release of red blood cells into the blood stream.
[J:310]

193 Hypovolaemic shock is characterised by a white, cold, sweaty skin, a rapid pulse and low blood pressure, rapid respiration, intense thirst and possibly restlessness and torpor.
[G:478; I:419]

194 Rapid transfusion of whole blood is required following haemorrhage. Saline alone is of relatively little value because it rapidly dis-

tributes through the entire extracellular fluid. Plasma expanders which remain in the cardiovascular space are of some use in maintaining the circulation. However, serum albumin tends to dehydrate the tissues by drawing out fluid. If restoring the blood volume to normal does not result in rapid restoration to health, then the patient is said to be suffering from irreversible shock—in which there has been some dispute as to whether vasoconstrictor drugs should be used to produce an increase in blood pressure or vasodilator drugs which encourage tissue perfusion.
[B:3-256; E:368; I:431]

The respiratory system

195 Respiration allows control of the P_{O_2}, P_{CO_2} and pH of the fluids of the body.
[C:497; H:1672; J:327]

196 The processes carried out by the respiratory system are exchange of air between the surroundings and the alveoli, exchange of the gases (O_2 and CO_2) between the alveolar air and the lung capillaries, transport of gases by the blood and their exchange between the blood and the tissues of the body.
[A:202; E:516; I:435]

197 The vital capacity (4.8 l in men, 3.1 l in women) is the maximum quantity of air that can be inspired following a maximum expiration. Tidal volume (0.5 l) is the amount of air moved out of the lungs at each expiration. The inspiratory reserve volume (3.3 l, 1.9 l) is that volume of air over and above the tidal volume inspired with a maximum inspiration. The expiratory reserve volume (1.0 l, 0.7 l) is that volume of air expired by an active expiratory effort after the tidal volume has been expelled. That volume of air remaining in the lungs after a maximum voluntary expiration is the residual volume (1.2 l, 1.1 l). Finally, the functional residual capacity is that volume of gas remaining in the chest after normal passive expiration (2.2 l, 1.1 l).
[D:68; E:521; F:166; H:1684]

198 The lung volumes may be measured using a spirometer. Residual volume is determined from dilution of a gas, such as helium, in the lungs.
[A:207; B:6-8; E:522; G:277]

199 Hyperpnoea is deep breathing, dyspnoea is rapid shallow breathing of which the patient is conscious and apnoea is a period of respiratory arrest.
[C:527; D:97; H:1682]

200 Air passes into the pharynx, a passage common to both air and food. The pharynx then branches into the oesophagus, which passes to the stomach, and the larynx, which passes to the lungs. Food going 'the wrong way' meets the vocal cords, closure of which prevents the entry of food into the lungs.
[E:526; F:164; J:361]

201 Pulmonary ventilation is the amount of air inspired per minute. Functionally, ventilation can be divided into alveolar ventilation, which is the volume of air reaching the alveoli where respiratory gas exchange occurs, and dead space ventilation, that volume of air not exchanged with the blood.
[B:6-10; D:64; H:1702; I:471]

202 The anatomical dead space comprises the volume of the air passages proximal to the respiratory bronchioles. The total or physiological dead space comprises the anatomical dead space plus the alveolar dead space—which is the volume in that portion of the lung containing poorly functioning alveoli. Neither the alveolar nor the physiological dead space has morphological correlates.
[B:6-11; E:525; F:167; H:1702; J:340]

203 The volume of the snorkel tube is approximately $128 \, ml - \pi r^2 l$, where r is the radius of the snorkel tube and l the length; i.e. $^{22}/_7 \times (0.95)^2 \times 45 \, ml$. An alveolar ventilation of 30 l·min^{-1} at 15 breaths·min^{-1} represents an alveolar ventilation of 200 ml. The tidal volume = alveolar ventilation + dead space = $200 + 140 + 128 \, ml = 468 \, ml$.
[C:504; G:279]

204 Air is warmed, moistened and cleansed as it passes through the dead space. Dust can be filtered out through the nasal hairs. The ciliated and mucus-secreting cells lining the dead space also help to clear airways of fine particulate matter.
[A:202; G:272]

205 Resistance lies in medium-sized bronchioles. One might expect it to lie in the small bronchioles, but there are so many arranged in parallel that the combined resistance is small.
[A:206; I:444]

206 In the upright posture ventilation per unit volume is greater at the bottom of the lungs, and in the supine position ventilation is greater in the posterior part of the lungs. This results from the distortion of the lung produced by its weight. In the upright position the intrapleural pressure is less negative at the base than the apex and so is more compressed in the upright state but expands more in the upright posture.
[A:216; G:281]

207 The alveolar PO_2 and PCO_2 change relatively little during expiration and inspiration. Because the functional residual capacity is 2 l, movement of 340 ml air has relatively little effect.
[C:506; J:345]

208 The composition of the inspired air is 21 per cent O_2 and 79 per cent N_2, and that of expired air 16 per cent O_2, 4 per cent CO_2 and 80 per cent N_2. The oxygen uptake is therefore

$$V_{exp} \left[\frac{80}{79} \times 21 - 16 \right] \%$$

where V_{exp} is the volume of the expired air. This becomes $5.27\% \times V_{exp} = 263.5$ ml·min^{-1}. Similarly, the CO_2 evolved $= 4\% V_{exp}$, which is 200 ml·min^{-1}.
[D:63; F:168]

209 The diaphragm and the internal intercostals of the parasternal region are active during quiet inspiration. The principal accessory muscles are the anterior neck muscles (the scalene and the sternomastoid muscles) which are used during forced inspiration and the abdominal muscles employed in forced expiration.
[A:205; B:6-36; E:517; F:158]

210 Expiration is a purely passive process except during the early phases when some inspiratory contraction persists.
[B:6-37; D:64; I:436]

211 Because of the pull of the lungs, the intrapleural pressure is below that of the atmosphere. On inspiration this becomes 5–10 mmHg (0.6–1.3 kPa) below atmospheric pressure. The difference between intrapleural and intra-alveolar pressure (approximately atmospheric) increases, the lung wall is pushed out, the pressure in the alveoli drops to -1 mmHg and air is sucked into the lungs. On expiration, as the thorax and lungs spring back to the original size, the intrapleural pressure is 3–5 mmHg (0.4–0.7 kPa) below atmospheric. The alveolar air is compressed, so the pressure becomes approximately 1 mmHg above atmospheric.
[B:6-45; C:498; H:1684; I:444]

212 An intrapleural pressure of 110 mmHg (15 kPa) may be voluntarily achieved during Valsalva's manoeuvre and one of 300 mmHg (40 kPa) on coughing.
[G:274; F:160]

213 The normal position of the lungs is achieved at rest by a balance between the elastic forces of the lungs and the chest wall.
[I:435; J:332]

214 Compliance defines the relation between the distending pressure (P) and the lung volume (V) and is given by V/P. In the example given, the distending pressure is 5 cmH₂O, so the compliance is 0.5/5 or 0.1 $l \cdot cmH_2O^{-1}$ or 1.0 $l \cdot kPa^{-1}$.
[G:275; H:1687; I:438]

215 For a sphere, the pressure $P = 2T/R$, where T is the tension in the walls and R is its radius. Because alveoli vary in size by some three to four times, the pressure required to keep them inflated would also vary. There is obviously a uniform pressure throughout the alveoli—a situation which is possible because of the surfactant which lines them.
[B:6-39; C:503; D:223]

216 Surfactant is vital for the first breath of the neonate, and in hyaline membrane disease of babies a deficiency results in a tendency for lungs to collapse and for reinflation to be difficult. The presence of surfactant is also important in the adult; in pulmonary embolism, surfactant is deficient in the wedge of tissue beyond the obstruction and there is thus a tendency for the lungs in that region to collapse.
[A:204; H:1681; J:342]

217 Compliance should be measured when there is no air flow, i.e. when the only determinant of pressure is the recoil of the lung. The pressure gradient is the difference between the pressure in the alveolar space and that in the pleural space. If respiration is stopped with the glottis open, the former is equal to the barometric pressure. The latter is given by the pressure in the oesophagus, which is in essence a flaccid tube exposed to pleural pressure. If inflation is carried out in steps, then a plot of transpulmonary pressure against volume is obtained, the slope of which gives lung compliance.
[F:165; I:438]

218 Compliance is decreased in pulmonary congestion, increased in emphysema and unaffected in asthma.
[C:503; E:524; G:275]

219 The fact that this plot is a hysteresis loop rather than a straight line results from the frictional resistance to air movement in respiration.
[C:503; E:520; G:275]

220 The work of respiration includes the effort in overcoming the elasticity of the lungs and chest and the frictional resistance in the tissues. The work of breathing is increased in diseases such as emphysema, asthma and congestive heart failure.
[A:206; B:6-53; J:341]

221 In determining the forced expiratory volume the patient exhales as rapidly and as fully as possible. The volumes expired at the end of 1, 2 and 3 seconds are determined and expressed as a percentage of the total volume expired. A normal subject will expire over 86 per cent in the first second, but the value is greatly reduced in

some forms of lung disease. In patients with partial airway obstruction it gives a better indication of the severity of the disease than the vital capacity itself.
[G:279; I:450]

222 The hydrostatic pressure difference between the apex and the base of the lungs has an important effect on the distribution of blood because the capillary bed is surrounded by the alveoli in which the pressure is nearly atmospheric. The intravascular pressure at the base of the lungs is 16 mmHg (2.1 kPa)—the pressure generated by the right ventricle + 11 mmHg (1.5 kPa)—so the vessels are distended, offering little resistance. The intravascular pressure in the upper part of the lungs is $16 - 11$ mmHg, or 5 mmHg (0.7 kPa), and the blood flow in this region is reduced.
[A:215; E:314; I:454]

223 Both the smooth muscles of the bronchioles and of the pulmonary arteries are sensitive to the local gas tensions. A low PCO_2 resulting from a poor blood supply results in constriction of the bronchioles. Conversely, a high PCO_2 resulting from poor ventilation produces bronchiolar dilatation. With regard to the arteries, a low PO_2 resulting from poor ventilation produces vaso-constriction and a high PO_2 in well ventilated alveoli results in vasodilatation. By these local mechanisms, ventilation and blood supply are balanced.
[C:507; E:314; J338]

224 The effect on the pulmonary arterioles of an increased oxygen tension in producing vasodila-tation is opposite to that in the systemic arterioles.
[D:43; J:388]

225 Oxygenated blood in the systemic arteries is diluted by blood which has supplied the lung tissue and the coronary arteries. There is also variation in the pulmonary ventilation/perfusion ratio.
[E:544; F:169; G:283; H:1715]

226 The diffusing capacity of the lungs is defined as the quantity of gas transferred each minute for each mmHg difference in the partial pressure of

the gas in the alveolar air and in the pulmonary capillary blood. It is dependent on the nature of the gas under consideration, the solubility in the alveolar membrane and the ease with which it diffuses through it. It also depends on the surface area of the lungs and the average thickness of the alveolar membrane.
[B:6-26; C:506; E:539; H:1709; I:468]

227 The increased oxygen diffusion depends on the increased number of open capillaries and dilatation of these capillaries. There is also a greater oxygen gradient.
[C:506; E:540; H:1711; I:714]

228 The carbon dioxide content may be determined by exposing the blood to a vacuum; the evolved CO_2 may then be absorbed, the reduction in the volume of gas giving the CO_2 content. To determine the oxygen content, the blood is first haemolysed, the haemoglobin converted to methaemoglobin and the mixture exposed to a vacuum. The carbon dioxide is first absorbed and then the oxygen, using sodium hydrosulphite.
[A:222; D:86; H:1724]

229 Oxygen and carbon dioxide electrodes are available to measure the gas tensions directly.
[A:222; B:6-18; D:86]

230 Dissociation curves for the respiratory gases give the carrying power of the blood for any particular gas. The O_2-haemoglobin dissociation curve is a plot of the percentage saturation of haemoglobin against the partial pressure of oxygen; that for CO_2 is the volume of CO_2 absorbed by 100 vol. blood. Such curves are plotted by exposing blood to different partial pressures of oxygen or carbon dioxide in a tonometer and then determining the content of the particular gas in the blood.
[F:172; G:100]

231 The oxygen capacity is the quantity of oxygen carried by 100 ml blood when fully saturated, whereas the oxygen content is that volume carried in the blood at a given tension.
[A:222; D:82]

232 The oxygen dissociation curve has a sigmoid form. The haemoglobin molecule, which may be represented as Hb, reacts with four molecules of oxygen to form HbO_8. When the haemoglobin has taken up small quantities of oxygen, the uptake of additional oxygen is favoured, giving the characteristic curve.
[C:511; D:82]

233 The fact that oxygen carriage is little affected at partial pressures over 60 mmHg (8 kPa) means that a nearly constant arterial oxygen is maintained even with a variable PO_2 in the alveoli. The large drop in percentage saturation between 10 and 50 mmHg (1.3 and 6.7 kPa) means that oxygen release can occur with a small change in capillary oxygen tension.
[F:172; H:1725; I:463]

234 A number of factors, including pH, PCO_2, temperature and carbon monoxide, affect the dissociation curve of oxyhaemoglobin.
[A:224; G:96; H:1727]

235 More oxygen would be given off at (*a*) a lower pH and (*b*) a higher PCO_2. Less oxygen would be given off (*c*) at the lower temperature. Carbon monoxide combines irreversibly with haemoglobin, so the oxygen-carrying capacity is reduced.
[B:6-15; C:511; E:549; H:1727; J:347]

236 The dissociation curve of myoglobin is a rectangular hyperbola rather than sigmoid. Myoglobin takes up only one molecule of oxygen and the curve lies to the left of the haemoglobin curve, so it can take up oxygen from haemoglobin and can release oxygen at the low PO_2 found in exercising muscles.
[A:224; C:513; H:1728]

237 The majority of carbon dioxide is carried in the blood as bicarbonate (42 ml in arterial and 44.8 ml in venous blood). Some carbon dioxide forms a neutral carbamino compound with amino groups of proteins, principally haemoglobin (3 ml in arterial and 3.7 ml in venous blood). A small amount also exists in simple solution (3.0 ml in arterial and 3.5 ml in venous blood).
[B:6-16; D:84; E:533; H:1730; J:350]

238 In addition to carbon dioxide being carried in solution and as carbamino-Hb in the red cells, they contain carbonic anhydrase which allows rapid hydration of carbon dioxide to carbonic acid. The bicarbonate formed diffuses into the plasma and the hydrogen is buffered primarily by the haemoglobin because deoxygenated haemoglobin binds more hydrogen ions than the oxygenated form. The electrical neutrality of the cells is maintained by the inward diffusion of chloride ions.
[D:84; E:533; H:1731; J:380]

239 The level of carbon dioxide in the blood is more effective in controlling respiration than is that of oxygen.
[C:521; E:561; F:184]

240 There are two groups of receptors. First are the central chemoreceptors, lying superficially on the brain surface and influenced largely by changes in the composition of the cerebrospinal fluid, especially the hydrogen ion concentration. The hydrogen ion concentration follows changes in the carbon dioxide tension. Second are the peripheral arterial chemoreceptors, located at the bifurcation of the common carotid artery and in the aortic bodies of the arch of the aorta. The cells of these chemoreceptors have a rich network of sinusoidal blood vessels and are sensitive to changes in the composition of the blood, especially of oxygen.
[A:232; E:566; G:289; H:1751, 1755]

241 The combination of carbon monoxide with haemoglobin reduces the amount of oxygen carried by the haemoglobin but not that oxygen (0.3 ml in arterial blood and 0.13 ml in venous blood) carried in solution, and it is thought that it is to this dissolved oxygen that the chemoreceptors respond. However, recent experiments have cast doubt on this idea.
[A:237; B:6-15; H:1761; J:356]

242 Chemoreceptor activity may also influence the vasomotor centre.
[C:460; D:43; H:1067]

243 High concentrations of carbon dioxide stimulate and low concentrations depress the respiratory centre.
[D:96]

244 Additional afferent pathways for the reflex control of respiration include: the vagal fibres from the lungs; afferent fibres from proprioreceptors in muscles, tendons and joints; afferent fibres from various receptors in the skin and membranes of the respiratory tract.

There are also inputs from various parts of the central nervous system to integrate respiration into actions such as yawning, coughing, sneezing and swallowing.
[B:6-57; D:96; G:297; H:1767]

245 Respiration is believed to be controlled by the inspiratory and expiratory centres in the medulla and apneustic and pneumotaxic centres in the pons.
[A:228; C:517; E:552; H:1767]

246 Inspiration is initiated by increased activity of the inspiratory centre which is dominated by the apneustic centre. The inspiratory centre also sends impulses to the pneumotaxic centre. In turn, impulses from this second centre, together with those from the pulmonary stretch receptors, inhibit the respiratory centre via the apneustic centre. Expiration commences and the subsequent reduction in the activity of the pneumotaxic centre, together with that of the pulmonary stretch receptors, finally results in the start of another respiratory cycle.
[B:6-58; E:557; G:288; H:1767; I:481]

247 Ideally, the ventilation and perfusion for any area of the lung are uniform and matched; the ratio of ventilation (in $l \cdot min^{-1}$) to perfusion (in $l \cdot min^{-1}$) is normally 0.8. Disease processes may prevent this, and so, for example, certain alveoli may not be ventilated which results in arterial hypoxia.
[A:218; B:6-33; C:530; H:1714]

248 Because the oxygen content of inspired air is generally constant at 20.19 per cent, hyperventilation can increase the arterial oxygen content very little over the normal value whatever the ventilation, whereas carbon dioxide content can be readily altered. In addition, the shape of the carbon dioxide dissociation curve favours compensation for $PaCO_2$ with very little increase in total ventilation.
[C:530; H:1843; I:477]

249 Hypoxia is a deficiency of oxygen in the tissues, hypoxaemia a reduction in the oxygen content of the blood and cyanosis a bluish discolouration of the tissues resulting from insufficient oxygenation of the blood in the lungs.

The four traditional categories of hypoxia are: hypoxic hypoxia where the PaO_2 is reduced; anaemic anoxia in which there is lowered haemoglobin; stagnant or ischaemic anoxia in which blood flow to the tissues is low; and histotoxic anoxia in which the cells are unable to make use of the oxygen they receive.
[G:305; J:363]

250 Other possible causes of arterial hypoxia are an anatomical (right-to-left) shunt, hypoventilation or an impaired diffusion across the alveolar membrane, but this last need not be considered in this context.

If the $PaCO_2$ is high, alveolar ventilation accounts for at least part of the hypoxia. If the $PaCO_2$ is normal or low then the cause is anatomical shunt and/or ventilation/perfusion mismatch. If, when 60 per cent oxygen is given, the PaO_2 rises above 100 mmHg (13.3 kPa) then the last is the only important cause of hypoxia.
[A:237; C:528; G:341; H:1849]

251 First it should be ascertained that there is both cardiac arrest and failure of respiration. Then the airways should be cleared and maintained clear. The circulation of the blood must be assisted by carrying out external cardiac massage, when intermittent pressure is applied with the heel of the hand over the lower end of the sternum to depress the sternum by 3–4 cm. Oxygenation of the blood must be carried out. If there is no respirator available, mouth to mouth respiration must be employed. Finally, an attempt must be made to get the heart beating.
[B:6-89; D:75; G:308; H:1833]

252 After 2–3 days' exposure to 100 per cent oxygen, pulmonary oedema may develop. There is also a tendency to atelectasis. Patients whose chief respiratory drive is hypoxia may also develop dangerous carbon dioxide retention and acidosis.
[D:74; G:308]

253 The most important functions of the kidney are probably regulation of salt and water balance and of pH and excretion of waste and of substances which are not metabolised. The kidneys also serve an endocrine function.
[B:5-28; E:438; H:1165; I:489]

254 About 1200 ml per day is drunk and about 1000 ml taken in with the food. In contrast, 1500 ml per day is lost in the urine.
[A:264; E:424; F:227]

255 The nephron originates as a blind sac, known as Bowman's capsule, which is the site of filtration. This opens into the proximal tubule where most of the solutes required by the body are reabsorbed and 75 per cent of the filtered water. Next is Henle's loop, which is associated with the development of a concentration gradient across the kidney. Finally the distal convoluted tubule and collecting duct are reached, where final adjustments are made to urine composition.
[B:5-24; D:134; E:439; J:370]

256 The cortical nephrons are those with glomeruli in the outer part of the cortex and they have short loops of Henle, whereas those with glomeruli in the juxtaglomerular region of the cortex have long loops which extend down to the medullary pyramids. In the human, nephrons with long loops represent only 15 per cent of the total.
[A:246; E:439; I:491]

257 The diagram should indicate the positions of the interlobular artery and veins, the afferent arterioles, the glomerular capillaries, the efferent arterioles, the peritubular capillaries and the vasa recta.
[C:539; E:440; G:132]

258 The glomerular filtrate is an ultrafiltrate of plasma and hence has the same composition as plasma save that it contains no components with a molecular weight greater than 68 000. The

major constituents of urine and plasma are shown below:

Substance	Urine	Plasma
Glucose	0.0 mmol·l^{-1}	5.5 mmol·l^{-1}
Sodium	30.0 mmol·l^{-1}	145 mmol·l^{-1}
Urea	1.8 mmol·l^{-1}	5.0 mmol·l^{-1}
Creatinine	196 mmol·l^{-1}	1.0 mmol·l^{-1}

[B:5-29; C:538; D:135; E:443; F:215]

259   The glomerular filtration rate is the volume of blood filtered as the blood flows through the glomeruli, and in man is 120 ml·min^{-1}.
[C:542; E:443; H:1173; J:371]

260   The filtration force $= (P_1 - P_2) - (\pi_1 - \pi_2)$, where P_1 is the glomerular capillary pressure, P_2 the normal intratubular pressure, π_1 the osmotic pressure of the plasma proteins and π_2 the oncotic pressure of the glomerular filtrate. It is thus $(60 - 10) - (30 - 1) = 21$ mmHg.
[E:444; I:500; J:372]

261   Lowered oncotic pressure of plasma would result in increased glomerular filtration, and reduced blood pressure and urinary obstruction in reduced filtration.
[A:246; E:444; I:564; J:372]

262   Constriction of the afferent arterioles reduces blood flow to the glomerulus and, thus, the glomerular pressure, which results in decreased glomerular filtration. Efferent arteriolar constriction, on the other hand, increases resistance to outflow from the glomeruli and, hence, raised glomerular pressure; this usually results in an increase in glomerular filtration rate.
[E:444; I:506]

263   Clearance is a mathematical concept and is defined as the volume of blood which is completely cleared of a substrate in 1 minute. An equation for calculation of the clearance (C) may be obtained from the basic statement that the amount of substance (S) appearing in the urine is equal to the amount leaving the blood. If U is the urinary concentration of S, and V the volume of the urine, then the amount of S appearing in the urine is UV. If the concentration of S in the

plasma is P, then the amount cleared from the plasma is CP. $UV = CP$; hence $C = UV/P$.
[B:5-34; C:143; E:453; F:228; H:1174]

264 Inulin clearance may be used as a measure of glomerular filtration rate.
[A:248; E:453; G:137; H:1173]

265 For a substance to be used to measure glomerular filtration rate, it must be freely filtered but have no effect on the filtration rate. It should be neither reabsorbed nor excreted by the tubules, and neither metabolised nor stored in the kidney. It obviously should not be toxic, but should be relatively easy to measure.
[C:452; E:453; J:373]

266 From the equation derived in answer 263, clearance (C) is given as:

$$C = \frac{18 \times 1.04}{0.3} = 62.4 \ \mathrm{ml \cdot min^{-1}}.$$

Creatinine is generally used in the determination of glomerular filtration rate.
[D:144; E:452; G:137]

267 Studies on the isolated perfused nephron *in vitro* have contributed greatly to our understanding of renal function. These techniques include continuous microperfusion, stopped flow (stationary) microperfusion and free flow micropuncture.
[A:257; F:215]

268 Excretion describes the appearance of a substance in the final urine, whereas secretion is the transfer of a substance from the peritubular plasma to the tubular lumen. Movement in the opposite direction results in absorption. Examples of substances which are secreted are p-aminohippuric acid (PAH) and penicillin. If the clearance of a substance is greater than inulin, it is secreted; if less, then it is reabsorbed.
[E:445; G:139; H:1179; J:370]

269 The clearance of a substance such as para-aminohippuric acid which is filtered and secreted by the tubules is equal to the renal plasma flow. If the haematocrit is known then the renal blood flow may be calculated.

As with other organs, the blood flow can be calculated using the Fick principle.
[A:249; B:5-43; E:453; F:228; H:1175]

270 Autoregulation is the process whereby the blood flow to an organ, in this case the kidney, is kept constant and independent of blood pressure over the range 80–200 mmHg (10.7–26–7 kPa). It may be brought about by local mechanism and by the release of renin.
[C:452; I:494]

271 Certain transport mechanisms can be inhibited both competitively and non-competitively and are dependent on a supply of energy, so they are defined as active.
[B:5-57; H:1179; I:514; J:378]

272 Active sodium transport in the kidney occurs with two sodium pumps. One pump extrudes sodium from the tubule lumen and the chloride moves passively with the sodium accompanied by water. The other pump exchanges a potassium ion for each sodium ion transported. Water uptake by the nephron results in relatively high concentrations of urea in the tubular urine and hence it diffuses into the capillaries.
[D:138; E:454; G:157]

273 It is postulated that glucose transport occurs through carriers in the renal tubule, so that there is a limited capacity for glucose transport. Glucose reabsorption is in proportion to the increase in plasma glucose till the transport maximum is reached and then no further glucose is reabsorbed despite the increase in plasma glucose. The tubular maximum (T_m) for glucose is about 320 mg·min^{-1}. Given that the GFR is 125 ml·min^{-1}, one would predict a value of 14 mmol·l^{-1} (320 mg·min^{-1}/125 ml·min^{-1}).

The actual value is 9.9 mmol·l^{-1} because not all the nephrons have the same T_m for glucose and GFR. In addition, some glucose is not reabsorbed because the processes involved in glucose transport are not completely reversible.
[B:5-37; D:145; F:229; H:1176]

274 The filtrate remains isotonic in the proximal tubule, is hypotonic at the beginning and isotonic at the end of the distal tubule, and

hypertonic in the end of the collecting duct, so urinary concentration effectively occurs in the collecting duct.
[A:251; B:5-50; E:457; I:526]

275 The concentrating mechanism depends on the hypertonicity of the inner medulla, which is due to sodium chloride and urea, and on the permeability of the collecting duct in certain circumstances to water.
[A:252; D:141; E:458]

276 The concentration gradient exists because the loops of Henle act as counter-current multipliers. While the descending loop of Henle is relatively impermeable to solute, it is permeable to water so that water leaves the tubule and the fluid becomes more concentrated. The thin ascending limb is permeable to sodium which therefore moves into the interstitium. The thick limb is impermeable to water and to solute, but chloride is actively transported out of this segment of the tube. Urea moves out of the collecting duct in the inner medulla and also contributes to the high osmolality. The gradient could be dissipated by the flow of blood through the kidney, but the vasa recta act as counter-current exchangers.
[A:250; C:551; E:458; H:1180; I:522]

277 Filtered L-amino acids are almost completely reabsorbed in the proximal tubule. Without reabsorption of essential amino acids, protein malnutrition would occur.
[C:546; I:511]

278 The juxtaglomerular apparatus comprises a complex of cells from the afferent and efferent arterioles and the macula densa segment of the early distal tubule. It contains renin, which is an enzyme acting on an α-globulin (angiotensinogen) in the plasma to produce angiotensin I which is subsequently converted to angiotensin II, a powerful vasoconstrictor agent.
[E:458; F:214; G:133]

279 Antidiuretic hormone (vasopressin) acts on the final section of the distal tubule and on the collecting duct, to increase the permeability of the tubule to water and hence produce antidiuresis. Aldosterone increases the tubular

reabsorption of sodium and excretion of potassium and hydrogen ions. Parathyroid hormone is thought to produce a decrease in calcium excretion through its action on the distal tubule. It also acts on the proximal tubule to reduce the absorption of phosphate. Glucocorticoids may also influence sodium excretion.
[A:252; B:5-61; H:1182; I:551]

280 Bicarbonate behaves as though it had a T_m of 28 $mmol \cdot l^{-1}$, being totally absorbed below this concentration. The enzyme carbonic anhydrase is present in the renal tubule cell which catalyses the formation of bicarbonate. The bicarbonate is reabsorbed together with a sodium ion which is exchanged for hydrogen ion. Hydrogen ions are excreted in the form of sodium dihydrogen phosphate or ammonium chloride, the ammonia being derived from glutamine and other amino acids in the kidney itself. Bicarbonate is reabsorbed in the proximal tubule, although there exists an acid-secreting mechanism in the distal tubule.
[D:138; E:491; G:157; H:1197]

281 Two types of diuresis exist: water diuresis, which is seen in the virtual absence of vasopressin, when very little water is reabsorbed; and solute diuresis, brought about by the presence of large quantities of solute in the renal tubule. Most diuretics produce an osmotic diuresis, altering the rate of excretion of sodium, potassium or chloride.
[C:554; E:512; I:548]

282 Such drugs are secreted into the proximal tubule by active transport processes and escape, to varying degrees depending on the urinary pH, via non-ionic diffusion from more distal parts of the nephron.
[C:548; I:516]

283 Pregnancy is normally associated with hypervolaemia and a marked increase in the renal plasma flow and glomerular filtration rate which is at its greatest in the second trimester. The filtered load of glucose is increased and glycosuria is common.
[E:1113; F:548]

284 In acute renal failure the kidney virtually stops functioning, as opposed to chronic renal failure in which the nephrons are progressively destroyed. Two of the commonest causes of acute renal failure are acute glomerulonephritis, a disease which results from an immune reaction in which the glomeruli are inflamed or even destroyed, and acute damage or obstruction of the tubules.
[E:504; F:234]

285 Chronic renal failure occurs when 75 per cent or more of the nephrons do not function. When the glomerular filtration rate falls to 50 per cent of normal, the blood urea concentration tends to rise, so the urea load for each nephron that is functioning is increased. An osmotic diuresis then ensues which results in impaired reabsorption of sodium and water. This in turn leads to loss of the diurnal rhythm of salt excretion and to nocturia.
[A:259; B:5-93; C:560; E:505; H:1206]

286 In dialysis, unwanted substances are removed by allowing them to cross a porous membrane into the dialysing fluid. This may be achieved by peritoneal dialysis, not requiring additional equipment, or with an artificial kidney in which blood is passed between two sheets of cellophane with the dialysing fluid on the outside. The membrane allows all constituents of plasma save protein to diffuse freely in both directions.
[D:147; E:507; F:235]

287 Any damage to or disease of the kidney which results in a fall in the vascular supply or in glomerular filtration rate results in hypertension. These events lead to retention of salt and water and eventually to hypertension.
[A:259; E:507]

288 On an appropriate diet which includes low protein and low sodium, life—albeit restricted—may be maintained if as little as 2 per cent of the nephron population is functioning.
[D:147; F:234]

289 Micturition involves the detrusor muscle, supplied by the sacral parasympathetic (+) and the sympathetic (−), possibly the internal

sphincter supplied by the parasympathetic (−) and the sympathetic (+), and the internal sphincter under voluntary control.
[E:502; G:149; H:912; J:372]

290 The law of Laplace states that the pressure in a sphere is proportional to the wall tension divided by the radius. Thus the bladder can accommodate increasing amounts of urine with only a small increase in the pressure within the bladder, until the volume is about 400 ml.
[B:5-97; C:562; G:150]

Digestion

291 The six essential components of an adequate diet are carbohydrates, fats, proteins, vitamins, minerals and water.
[A:18; C:362]

292 The vitamins are as follows:

Fat-soluble vitamins	Examples of sources
A	Butter, egg yolks, carrots
D_2 and D_3	Butter, eggs
E	Cereals, liver, eggs
K	Vegetables (e.g. cabbage, cauliflower)
Water-soluble vitamins	
B group	Yeast, meat
C	Fresh fruits, citrus fruits

[A:23; C:240]

293 Vitamin A deficiency leads to failure of growth, loss of weight, decreased resistance to infection and keratinisation of the epithelium of the eye and respiratory tract. Lack of thiamine leads to beri beri; of riboflavin to disturbances of the skin, cornea and mouth; and of ascorbic acid to disturbances of calcification of bone and formation of enamel in teeth. Rickets results from lack of vitamin D.
[D:123; E:979; F:473]

294 The important inorganic substances in the diet are calcium, sodium, potassium, magnesium, iron, phosphorus, iodine and chloride.
[D:128; G:344]

295   The daily requirement of calcium is about 0.8 mg
and of iron 12.0 mg. Calcium is important for a
great number of functions, including the
permeability of membranes, function of nerve
and muscle, clotting of blood and bone
formation. Deficiency of iron is associated with
anaemia.
[A:22; F:45, 250]

296   The equations are:
After food intake: energy intake = external activity
+ heat
In the fasting state: energy released in catabolism
= external activity + heat + energy stored.
[C:207; E:949; I:663; J:463]

297   The basal metabolic rate is a measure of the
energy requirement under basal conditions and
is generally expressed in terms of the surface
area of the body. Ideally, it should be measured
when the subject is at rest and 12–18 hours after a
meal, the mean value in a man being 170
$kJ \cdot m^{-2} \cdot h^{-1}$ (40 $kcal \cdot m^{-2} \cdot h^{-1}$) and that in a woman
150 $kJ \cdot m^{-2} \cdot h^{-1}$ (37 $kcal \cdot m^{-2} \cdot h^{-1}$).
[E:953; F:204; G:329; H:1359]

298   A bomb calorimeter consists of a steel chamber in
which a measured amount of foodstuff is
combusted in an atmosphere of oxygen. The
quantity of heat derived is determined by
measuring the increase in temperature of a
known volume of water in which the calorimeter
is placed; 1 g of carbohydrate produces 17 kJ (4.1
kcal) 1 g of protein 22 kJ (5.3 kcal) and 1 g of fat 39
kJ (9.2 kcal).
[G:328; H:1352; I:665]

299   The respiratory quotient (RQ) =

$$\frac{CO_2 \text{ produced}}{\text{oxygen consumed}}$$

The RQ when carbohydrates, fats and proteins
are metabolised is 1.0, 0.7 and 0.8, respectively.
The RQ for a mixed diet is approximately 0.8.
[A:13; E:972; F:202]

300   At the start of exercise respiration is stimulated
and carbon dioxide is blown off. During the
course of exercise the lactic acid produced by the
muscles is buffered in the blood to form carbonic

acid which stimulates respiration and more carbon dioxide is given off, so the ratio of CO_2 produced/O_2 used is greater than 1.0.
[C:209; F:203; H:1358]

301 The kilojoule or calorie equivalent of oxygen is the energy production for a given volume of oxygen for a given RQ.
[D:120; E:953; G:327]

302 If an RQ of approximately 0.8 is assumed, then each litre of oxygen would produce a given amount of energy—in this case 20 kJ (4.8 kcal). Thus from oxygen usage the production of energy can be calculated. This value can be divided by the surface area of the body to give the metabolic rate.
[C:209; E:953; H:1359; J:463]

303 Metabolic rate can be determined by direct calorimetry using an Atwater–Benedict respiratory calorimeter.
[C:208; J:463]

304 A mean value for metabolic rate should not be calculated for a group of male and female subjects, as the value is affected by the sex of the subject. Metabolic rate is also influenced by age, starvation, body temperature and the circulating concentrations of thyroid hormones and adrenaline.
[A:16; E:951; F:204; H:1361]

305 Under basal conditions metabolism proceeds at a certain rate, but when food is taken there is an increase in the rate of metabolism. This is called the specific dynamic action. Its exact cause is unknown, but it is greatest after intake of protein.
[C:209; G:330]

306 There must be a system for regulating food intake and balancing it with metabolic activity, as in most individuals body weight remains relatively constant for years in spite of altering metabolic requirements.
[B:2-5; H:1374; J:466]

307 The biochemical and physiological basis of food intake is not well understood, but it is known that the hypothalamus plays an important role.

Afferent inputs from the gastrointestinal tract (e.g. contractions of the stomach as in hunger) and environmental temperature affect intake. The sense of taste may act as a guide to nutrition; for example, sodium-deficient sheep and rats choose to drink salt, rather than pure, water. Eating habits themselves are also important.
[E:974; H:1379; I:223; J:467]

308 The rate of oxygen consumption is 300 ml·min^{-1}; assuming an RQ of 0.8, this is equivalent to 5.9 kJ·min^{-1} (1.4 kcal·min^{-1}), given that 1 litre of oxygen is equivalent to 20.1 kJ (4.8 kcal). The surface area = 0.007184 × 50$^{0.425}$ × 162$^{0.725}$ = 1.43 m^2. The metabolic rate is thus 245.65 kJ·m^{-2}·h^{-1} (58.7 kcal·m^{-2}·h^{-1}).
[D:120; F:204]

309 Experiments have been performed in animals in which either lesions are placed in the hypothalamus or regions are stimulated electrically. There appears to be a satiety centre located in the midline of the hypothalamus, as stimulation of this region results in cessation of eating and ablation in overeating. In contrast, there seems to be an appetite centre in the outer hypothalamus, as stimulation induces eating and injury inhibits it.
[A:18; B:2-3; E:973; H:1375; I:223]

310 The four main activities of the gastrointestinal tract are secretion of digestive juices, digestion, absorption and, finally, motility.
[C:371; J:402]

311 Below the serosa are two layers of muscle: an outer longitudinal layer and an inner circular layer. A third layer of smooth muscle, the muscularis mucosae, lies between the submucosa and the mucosa. The latter contains most of the exocrine cells in the gut and the epithelial cells involved in absorption. Between the two outer layers of muscle is found a plexus of nerves, the myenteric or Auerbach's plexus, with a second plexus, the submucous or Meissner's plexus, below the circular muscle. The luminal surface of the tract is highly convoluted.
[C:371; E:405; J:405]

312 The activity of both the exocrine glands and the smooth muscle is controlled by the external autonomic nerves, internal nerve plexuses and hormones secreted by the gastrointestinal tract. The endocrine glands involved are controlled by these same inputs as well as by changes in the composition of the gastrointestinal contents. Similarly, there are nervous receptors sensitive to lumen contents as well as to wall distension. Digestion is also controlled in part from higher centres.
[B:2-13; C:371; J:405]

313 Stimulation of the sympathetic nerve supply decreases gastrointestinal motility, whereas stimulation of the parasympathetic increases it.
[C:160; I:188]

314 The salivary glands, of which there are three pairs in man, comprise secretory cells arranged in acini. The salivary enzymes are discharged into a system of small ducts and then larger ducts. There are two types of acinar cells and both secrete salt and water. The serous cells secrete in addition amylase, and the mucous cells a number of glycoproteins collectively called mucin. The cell type and nervous supply are indicated below

Gland	Acinar cell type	Nerve
Parotid	Serous	Glossopharyngeal
Sublingual	Mucous	Facial
Submaxillary	Mixed	Facial

[A:36; B:2-25; C:375; E:871; H:1289]

315 Saliva is generally hypertonic, the osmolality increasing with increasing flow.
[A:38; B:2-29; H:1289; I:267]

316 In general, the ratio of concentration in saliva to that in serum is less than 1.0 for Na^+ and Cl^-, and greater than 1.0 for K^+ and HCO_3^-.
[A:38; H:1291; I:267]

317 Sympathetic stimulation is chiefly associated with flow changes, and parasympathetic with changes in organic content.
[A:38; F:403; H:1292]

318 The main stimuli for the secretion of saliva are the presence of food in the mouth and the sight, smell or thought of food.
[B:2-32; G:313; H:1293; J:421]

319 Swallowing is a complex reflex. As the food is moved into the pharynx, the soft palate is raised, preventing the food from entering the nasal cavity. Respiration is inhibited and the glottis is closed, preventing food from entering the trachea. In addition, the bolus of food presses the epiglottis over the glottis.
[D:104; J:421]

320 The bolus of food is moved along the oesophagus by peristaltic waves.
[D:104; G:314]

321 This diagram should indicate the oesophagus, cardia, fundus and body of the stomach where the parietal and chief cells are found. The antrum, pylorus and duodenum should also be indicated.
[A:44; C:377; E:858]

322 The three classic phases of gastric secretion are cephalic, gastric and intestinal.
[B:2-49; D:107; E:874; F:106; H:1299]

323 The oxyntic cells secrete HCl, and the peptic cells secrete pepsinogen under the influence of vagal stimulation and the hormones gastrin, cholecystokinin-pancreozymin (CCK-PZ), secretin and enterogastrone.
[G:314; H:1296; J:423]

324 Hydrochloric acid, produced as a surprisingly strong solution to be found in a biological system, is approximately $0.1 \, mol \cdot l^{-1}$ and reduces the pH of ingested material to 1.5–2.5. Chloride and hydrogen ions are actively pumped into the gastric lumen. The source of hydrogen ions is not clear, but is associated with the production of bicarbonate which passes into the venous blood, producing the alkaline tide.
[A:54; B:2-42; C:378; E:872; H:1296]

325 The control of gastric acid secretion was studied in dogs using isolated stomach pouches in which the gastric secretory activity could be followed after various stimuli. Examples of such pouches

are the Pavlov pouch, which has the nerve supply intact, and the Heidenhain pouch, which has been denervated.
[D:106; G:314]

326 The approximate daily volume of gastric juice is 1–2 litres. The acid output of the stomach in man may be studied with a fractional test in which gastric secretion is stimulated with histamine or pentagastrin and the stomach aspirated every 15 minutes and the acid content of the samples analysed.
[B:2-41; F:407; I:629]

327 Gastrin is produced in the antrum and is released in response to food (both distension and digestion products) in the stomach and in response to vagal activity.
[A:56; B:2-21; D:107; E:873; G:315; H:1314]

328 In some patients the protective barrier may become weak and break up; acid diffuses back into the mucosa, causing strong contractions and pain and the secretion of histamine and pepsinogen. Histamine stimulates acid release and, together with other substances produced following mucosal damage, causes increased capillary permeability, and so the tissue becomes oedematous and exudes fluid. Plasma proteins are thus lost and bleeding may occur which can vary from minor to exsanguination.
[A:29; C:382]

329 Gastrectomised patients are prone to anaemia. Intrinsic factor present in the stomach is required for the absorption of vitamin B_{12}, and the gastric juices dissolve iron and can provide an environment for it to be reduced to the ferrous state, in which form iron is most easily absorbed. The 'dumping' syndrome can also be seen in these patients. Rapid absorption of glucose from the intestine results in prompt insulin release and hypoglycaemia ensues. Large quantities of hypertonic fluid reaching the intestine lead to movement of water into the gut and a subsequent fall in plasma volume.
[C:380; E:894; F:407]

330 The smooth muscle of the gut comprises small fibres organised in bundles and connected by

gap junctions so that current can flow between them. In contrast, the fibres of striated muscle are long and independent save for the fact that groups are controlled by the same motor neurone. The origin and insertion of smooth muscle is in connective tissue, whereas for striated muscle it is in bones. Smooth muscle is supplied by nerve plexuses and striated muscle has a motor end-plate. The resting potential of smooth muscle is variable, but that of striated muscle is not. Smooth muscle contraction and relaxation are slow, those of skeletal muscle rapid.
[G:56, 76; J:248]

331 The segmenting contractions are non-propulsive movements of which the function is mixing the intestinal contents. The peristaltic waves propel the food along the lumen.
[B:2-117; E:861; G:321; H:1337; I:618]

332 Powerful contractions develop in the empty stomach and as they become stronger 'pangs of hunger' are felt. When food enters the stomach, it relaxes (receptive relaxation) and thereafter the tone is gradually regained.
[B:2-8; E:859; G:317; H:1379; I:611]

333 The basic electrical rhythm is a wave of depolarisation originating in a pacemaker located in the longitudinal muscle layer in the upper stomach and spreading to the circular muscle. This depolarisation, also called the gastric slow wave, co-ordinates the peristaltic contractions.
[G:316; J:427]

334 Gastric emptying depends on the propulsive activity of the gastric muscle and the co-ordinated activity of the pyloric antrum, sphincter and duodenal cap. Antral contractions expel the food through the open sphincter, which contracts, followed by contraction of the duodenal cap. Gastric emptying is dependent on the volume of the stomach contents. Emptying is delayed if fat or solutions of low pH or high tonicity are present in the duodenum.
[A:46; E:859; H:1334; I:613]

335 The 'vomiting centre' in the reticular formation of the medulla may be stimulated following

anaesthesia and surgery, and, as the upper respiratory tract reflexes are lost with general anaesthesia, it is possible that the patient could inhale vomitus and suffocate.
[A:64; C:163]

336 The exocrine pancreas produces an aqueous juice characterised by a high concentration of bicarbonate and an enzyme-rich juice containing trypsin, chymotrypsin, amylase, maltase and lipase.
[E:876; G:319; H:1302; J:428]

337 The proteolytic enzymes are secreted in an inactive form, as, for example, trypsinogen and chymotrypsinogen. These forms are activated on intraluminal contact with enterokinase or by trypsin itself.
[A:66; B:2-58; E:876; F:409; H:1304]

338 A vagal reflex, causing pancreatic flow shortly after ingestion of food, results in secretion of a juice rich in enzymes. Secretin, which is released from the duodenal mucosa as a result of the presence of acid and to a lesser extent food products, stimulates the flow of large amounts of aqueous juice. Cholecystokinin, secreted as a result of digestion products in the duodenum, causes an increase in the secretion of enzymes.
[F:409; G:319; H:1304]

339 First, secretin and cholecystokinin-pancreozymin (CCK-PZ) potentiate each other's actions on pancreatic secretion. Secretin also inhibits gastric and intestinal motility, increases the flow of bile, stimulates insulin secretion and inhibits glucagon secretion. The hormone CCK-PZ, whose actions were originally believed to be due to two separate hormones, has an important action causing the gall bladder to contract—hence the name cholecystokinin. In general, agents which bring about emptying of the gall bladder are called cholecystagogues. In contrast to secretin, CCK-PZ stimulates gastric and intestinal motility. It also has weak acid-stimulating properties in the stomach.
[A:70; B:2-21; D:109]

340 Loss of alkaline pancreatic secretion leads to acidosis and may occur in diarrhoea.
[A:270; D:102]

341 The concentration of pancreatic enzymes in duodenal juice may be determined after administration of a test meal. Contamination with gastric juice may be prevented by aspirating the gastric contents or using a tube with inflatable balloons to isolate a region of the duodenum.

A more complicated test is to determine the volume of bicarbonate, enzyme and bilirubin concentration after intravenous secretin and pancreozymin.
[A:69; B:2-64; D:109]

342 Approximately 500 ml bile is secreted per day. It is an alkaline electrolyte secretion containing bile salts, bile pigments, cholesterol, fatty acids and lecithin.
[E:878; I:643; J:429]

343 The bile pigments biliverdin and bilirubin are breakdown products of haemoglobin and are present in the form of glucuronides. The bile salts are the sodium and potassium salts of glycocholic and taurocholic acids (bile acids). The bile acids are formed by the conjugation of glycine and taurine with cholic and related acids.
[F:412; G:321]

344 Choleretics are substances which increase the secretion of bile. Important in this respect are the bile acids. On the other hand, cholagogues are substances which cause contraction of the gall bladder, and example being CCK-PZ.
[B:2-77; D:109; I:645]

345 The total bile salt pool in the body is only 3.9 g and it has to be recycled twice during a meal to allow digestion of the fat. The bile salts are secreted by the liver and stored in the gall bladder between meals. Then the salts enter the intestinal tract, being reabsorbed passively in the jejunum and actively in the terminal ileum. Finally, the portal blood returns them to the liver. About 10 per cent of the bile salts is lost in the faeces with each cycle.
[A:101; B:2-70; E:939; G:321]

346 Digestion and absorption of fats is a complex process because ingested fats are insoluble in water. The first step is the formation of an

emulsion which exposes a greater surface area over which the pancreatic lipase can act. This enzyme splits the triglycerides to give fatty acids and β-monoglycerides which, with the bile salts, fat-soluble vitamins and cholesterol form micelles. It is thought that the micellar contents exchange with dispersed molecules and these latter are absorbed. In the microvilli, triglycerides are resynthesised and are subsequently coated with lipoprotein to form chylomicrons which are taken up by the lacteals.
[C:367; J:412]

347 Bile salts are important in the digestion of fat, up to 28 per cent of the ingested fat appearing in the faeces. Their importance lies in their ability to reduce surface tension so that, in conjunction with phospholipids and monoglycerides, they are responsible for the emulsification of fat and micelle formation. They also activate lipases in the intestine.
[A:73; D:109; E:883]

348 Plasma bilirubin is elevated in jaundice. Elevated concentrations may result from overproduction, as for example in haemolysis or impaired transit of bilirubin through the liver cells as a result of cell damage or specific defects in the metabolism. Impaired excretion in the bile canaliculae (intrahepatic cholestasis) or in the bile ducts (obstructive jaundice) may also result in jaundice.
[A:103; E:941; F:40]

349 In western countries 85 per cent of gallstones are cholesterol stones formed when the ratio of bile acids plus phospholipids to cholesterol becomes critically reduced. The cholesterol crystals thus formed gradually coalesce to form stones. No symptoms may be experienced if the stones remain in the gall bladder.
[B:2-83; C:387; D:11; E:941]

350 First of all the intestine is very long—some 6 m. The mucosa is thrown into folds and possesses villi which in turn have microvilli.
[A:71; E:885; H:1255; J:409]

351 It takes about 4 days for the cells of the intestinal mucosa to be completely replaced. The cells at the base of each villus are in active mitosis and

114

new cells move up into the villus, partly as a result of the volume of new cells and partly by active migration. Once they are sloughed off, they are digested in the lumen of the gut and much of the material is reabsorbed.
[C:389; j:416]

352 Cytotoxic drugs and ionising radiation produce their damaging effects on actively dividing tissue (e.g. malignant cells). They therefore affect the rapidly dividing cells of the intestinal mucosa and, hence, intestinal absorption.
[A:72; G:16]

353 Net absorption is the total absorption (absorption of both dietary and endogenous material) minus the loss of endogenous material in the faeces. The three steps involved in absorption are the uptake of substances from the lumen, their transport through the cells and, finally, their passage to the blood stream.
[I:648; J:411]

354 The three ways in which substances can pass into the mucosal cells are by simple diffusion, facilitated diffusion and active transport.
[A:74; B:2-89; E:887; H:1257; I:648]

355 The simplest method of studying absorption in man is to measure the blood concentrations after instilling a solution into the gut. Alternatively, studies may be carried out in which the difference is ascertained between the amount taken in and the faecal output. Recently, a system of tubes has been developed to allow reabsorption of a substance to be studied in a given segment of intestine. Preparations of animal tissue *in vitro* have also been used, the most successful being the everted sac.
[F:411; I:650]

356 Up to 10 litres per day of fluid may be reabsorbed. This is largely fluid added in the form of digestive juices with represent some four times the dietary intake. Most is absorbed in the small intestine where, in contrast to the stomach, pure water is rapidly absorbed.
[B:2-89; E:887; H:1259; I:688; J:414]

357 Both at rest and during digestion and absorption the contents of the colon and intestine are

isotonic with plasma. The duodenum plays a role in achieving osmotic equilibrium.
[H:1261; I:658]

358 The net movement of salt and water into or out of the lumen of the intestine results from bidirectional fluxes. At approximately 210 $mmol \cdot l^{-1}$, fluxes of both salt and water are in balance. Below 210 $mmol \cdot l^{-1}$ the outward flux of both salt and water is greater than the inward flux, resulting in net absorption. Above 210 $mmol \cdot l^{-1}$ there is relatively little movement of water, but flux into the lumen predominates. There is also net absorption of sodium.
[B:2-91; I:658; J:443]

359 Approximately 3 μg of vitamin B_{12} is required per day in the adult, a deficiency resulting in pernicious anaemia. Vitamin B_{12}, or cyanocobalamin, is a complex of four substituted pyrrole groups round a cobalt atom and has a molecular weight of 1353, so some mechanism is needed for absorption. This takes the form of intrinsic factor, a glycoprotein coming from the stomach. The B_{12} binds to intrinsic factor and the complex then attaches to a receptor in the brush border and B_{12} enters the absorptive mucosa.
[B:2-100; E:60; H:1284; I:637; J:414]

360 Iron is absorbed most readily in the ferrous form, although most dietary iron is in the ferric form. Ascorbic acid is important in achieving this reduction of ferric to ferrous iron. Iron absorption is an active process, occurring in the upper part of the small intestine.
[A:75; E:62; H:1282; J:416]

361 Diarrhoea is the passage of large quantities of soft or fluid stools. Excluding conditions associated with steatorrhoea, this results from impaired water and electrolyte absorption or rapid intestinal transport. Any solute such as magnesium sulphate which is not reabsorbed retains water within the intestine and induces diarrhoea. Lactose, which is present in the intestine as a result of lactase deficiency, has a similar effect.
[I:659; J:438]

362 Most of the enzymes are found in the epithelial cells of the intestine. Thus disaccharidases,

peptidases and enzymes involved in the breakdown of nucleic acids are located in the membrane of mucosal cells. Only enterokinase and intestinal amylase are thought to have any role in intraluminal digestion.
[D:109; F:411; H:1313]

363   It is generally accepted that carbohydrates are absorbed as monosaccharides. Pancreatic enzymes in the lumen of the intestine hydrolyse carbohydrates to disaccharides. Monosaccharides are formed on the surface of the brush border of epithelial cells.
[B:2-92; E:889; H:1264; I:649; J:411]

364   Some sugars, such as mannose and possibly fructose, are absorbed by diffusion. Others, such as glucose and galactose, are absorbed by an active process involving a carrier. Sodium is required for the latter process and one hypothesis is that sodium increases the affinity of the glucose molecule for the carrier and also aids transport of the carrier–monosaccharide complex into the cell.
[A:76; E:887; I:649]

365   Protein absorbed from the small intestine is derived from both the food taken in and from the desquamated cells and the enzymes secreted into the lumen of the gastrointestinal tract.
[H:1272; I:653; J:412]

366   Proteins and peptides are largely absorbed in the form of amino acids. Small amounts of protein are occasionally absorbed by the adult as witnessed by the fact that allergies may be developed against certain foodstuffs. In new-born mammals protein can be absorbed by pinocytosis which is important as antibodies can be taken up in this way.
[A:77; B:2-95; I:653]

367   There are, in the brush borders of mucosal cells, enzymes which hydrolyse peptides. The amino acids produced are transported into the mucosal cells by several different carrier systems.
[A:77; E:890; F:448; H:1274]

368   Because the small intestine has such a large capacity, limited resection has no adverse effect.

117

Extensive resection results in reduced absorption and hence steatorrhoea, large quantities of stools and malnutrition. Resection of the terminal ileum also results in a failure to absorb vitamin B_{12} and in steatorrhoea as the enterohepatic circulation of bile salts is interrupted.
[C:391]

369 The important functions of the colon are reabsorption of water and storage of dehydrated material.
[A:82; E:891; J:437]

370 The diagram should show the terminal ileum, ileocaecal valve, caecum, ascending colon, hepatic flexure, transferse colon, splenic flexure, descending colon, pelvic colon, rectum and the internal and external anal sphincters.
[C:391; F:418; H:1341]

371 The large intestine is inactive for much of the time. The major type of movement is the production of segmenting contractions (haustra). About three times a day, a mass movement occurs in which strong contractions of the proximal colon drive the contents into the descending and sigmoid colon. The haustra disappear at this time.
[A:79; B:2-123; J:435]

372 The contractions of the colon serve to delay rather than promote transit, so absorption of water is promoted. Thus intraluminal pressure is decreased in diarrhoea and increased in constipation, rather than the other way round.
[J:439]

373 In patients with complete spinal transection, distension of the rectum results in reflex defaecation brought about by contractions of the terminal colon and relaxation of the internal and external anal sphincters. In normal adults conscious control is exerted over this reflex and defaecation aided by contraction of the abdominal muscles and forced expiration against a closed glottis.
[A:81; C:393; E:865; H:1343]

374 Removal of the gland should produce symptoms associated with deficiency of secretion, which should be corrected by injection of an extract of the gland or transplantation of the gland to another site. Extracts of the gland should be prepared, the putative hormone purified, its composition identified and an assay established. Final proof comes from the demonstration that the concentration in the venous effluent is higher than in the artery.
[B:7-3; F:483; I:728]

375 The activities of the endocrine and nervous systems are integrated. For example, the anterior pituitary whose hormones control many endocrine glands in the body is itself controlled by the hypothalamus. The posterior pituitary is itself a neurosecretory system and the adrenal medulla is directly innervated.
[I:731; J:191]

376 Some peptide hormones, such as insulin and parathyroid hormone, are first synthesised in the form of larger precursor molecules or prohormones from which are split off the active hormones. The prohormone may represent a storage form of the hormone.
[A:89; C:259]

377 Most hormones are either peptides or steroids. However, the thyroid hormones are iodine-containing amino acids and the catecholamines are derivatives of tyrosine.
[D:159; J:191]

378 Peptide hormones are synthesised in the pituitary, the parathyroid gland, the thyroid (calcitonin) and the endocrine pancreas. The steroid hormones are synthesised in the adrenal glands and the gonads.
[D:159; E:989; J:192]

379 Generally only one hormone is secreted by one cell. It was suggested that vasopressin and oxytocin were synthesised in the same cells, but this appears not to be the case. A more complex situation exists with respect to steroid hormones, where complex metabolic pathways exist with many intermediates being hormones.
[J:208]

380 Hormones are also produced by the gastrointestinal tract (see questions 327, 339), the placenta and the kidney.
[A:462, D:159]

381 The simplest mechanism of controlling hormone secretion is by feedback from the metabolite being regulated by the hormone. Examples are insulin secretion regulated by blood glucose concentrations and parathyroid hormone secretion by blood calcium.
[G:189; I:732; J:208]

382 The trophic hormones are those hormones produced by the anterior pituitary which control secretions of other endocrine glands. The diagram should indicate that many factors contribute to the control of their release. There is control from the hypothalamus by way of hormones or factors which reach the anterior pituitary via the portal blood system; the hormones from the target organs feed back at the level of the hypothalamus and pituitary, and in some instances a short loop has been suggested whereby the trophic hormone itself exerts a negative feedback at the level of the hypothalamus.
[D:161; H:1483; I:731]

383 Specificity is vested in the molecular structure of the hormone. Specific receptors are located in the cell membrane. As with the interaction between enzyme and substrate, so the lock and key principle is thought to operate whereby the hormone fits into the receptor. In addition, specificity lies within each cell so that each gives the response characteristic of that specific tissue.
[A:462; G:189]

384 A hormone is first synthesised and generally stored before release into the circulation which transports it to the target tissue. The hormone may then be activated at this site. It is degraded either at the receptor site or in tissue such as the liver and kidney, and the products are excreted.
[B:7-5; G:189; J:194]

385 Peptide hormones generally act through the second messenger, cyclic AMP. Alternatively, peptide hormones may affect permeability or

exert an influence directly at the nuclear level.
[A:462; B:7-7; I:729; J:136]

386 Neither growth hormone, nor prolactin nor insulin acts via cyclic AMP. Insulin and growth hormone affect permeability, whilst both growth hormone and prolactin have a direct effect on protein synthesis. Hormones seem to initiate a series of interdependent changes in cellular metabolism, the first being the primary event. Thus stimulation of amino acid transport could be the primary event and protein synthesis within the cell the secondary event.
[B:7-7; C:267; F:485]

387 The hormone first binds with the receptor in the cell membrane. This complex then interacts with the adenylate cyclase system located in the internal plasma membrane, causing an increase in catalytic activity and, in turn, to formation of cyclic AMP.
[A:462; B:7-8; C:18; E:990; H:1463]

388 The cyclic AMP formed in response to hormone action influences many functions, including enzyme activity and permeability, so that the response characteristic of the hormone is seen. Most of the varied effects of cyclic AMP reflect its ability to activate a large variety of kinase systems, kinases being the enzymes which catalyse the transfer of a phosphate from ATP to an acceptor. This last could be a molecule in the membrane, ribosomes or an enzyme.
[E:990; F:484; H:1464; J:136]

389 A hormone is said to exert a permissive effect if its presence is necessary for the full expression of the action of another hormone. Thyroxine and glucocorticoids have permissive effects—both enhancing, for example, the lipolytic effects of catecholamines. Glucocorticoids are also required for catecholamines to produce their pressor effect and bronchodilatation.
[C:288; G:189]

390 Steroids bind to receptors within the cell and produce their effects at the nuclear level. They are able to do this because they are relatively small and lipid-soluble.
[B:7-11; E:991; H:1465; J:199]

121

391 Thyroid hormones do cause mitochondrial and other changes within the cell but these may be secondary effects. In many of the experiments performed, very high doses of the hormone were used. The thyroid hormones are thought to promote enzyme synthesis and, hence, metabolic effects.
[C:251; I:775]

392 A bioassay is a means of measuring hormone concentration in a given fluid. The response to standard and unknown are compared using the intact animal or isolated tissue. The response may be manifested immediately as with oxytocin (milk ejection) or angiotensin II (increase in blood pressure), or may take several days (e.g. the effect of growth hormone on tibial growth or prolactin on mammary tissue growth *in vitro*).
[E:989; F:484; H:1461; I:730]

393 The basic reagents for a radioimmunoassay are antibody and standard and labelled hormone. In this assay the labelled hormone reacts with the antibody and this antibody-labelled hormone complex can then be separated from the free or unbound labelled hormone because of its different physical properties. If a standard preparation of hormone or a solution containing an unknown amount of hormone is now introduced, it will compete with the labelled hormone for binding sites and will displace the label from the antibody; the greater the amount of unlabelled hormone, the greater the amount of label displaced. Thus the amount of unknown present can be determined by the amount of label displaced, because a given amount of standard solution displaces a given amount of label. The radioimmunoassay is much simpler to perform than the bioassay and allows processing of large numbers of samples. However, the amino acid sequence of the hormone representing the antigenic determinant is not necessarily the sequence conferring biological activity, so the results of immunoassay may not give the amount of biologically active hormone.
[B:7-6; E:989; H:1462; I:730]

394 The posterior pituitary comprises nerve endings of axons from the supraoptic and paraventricular nuclei, neuroglial cells and pituicytes. The blood supply is derived from the posterior or inferior

hypophyseal arteries. The anterior pituitary comprises cords of granular cells and an extensive network of sinusoids, the blood supply coming from the hypophyseal arteries and the hypothalamo-hypophyseal system. The posterior pituitary derives from the neural ectoderm as a downgrowth of the floor of the third ventricle, whereas the anterior pituitary derives from the buccal ectoderm.
[A:463; C:311]

395 The hormones synthesised by the anterior pituitary include growth hormone and prolactin produced by the acidophils—the somatotrophs and the mammotrophs respectively. The trophic hormones ACTH, FSH, LH and TSH are produced by basophils in the anterior pituitary.
[B:7-13; C:311; E:993; I:737]

396 When principles active in stimulating anterior pituitary hormone release were first postulated they were termed 'factors'. When these principles were isolated and their chemical structure identified, they were given the name 'hormone'.
[A:406; C:268; H:1480; J:205]

397 If the anterior pituitary is replaced next to the severed pituitary stalk, there is some hypothalamic control of anterior pituitary hormone secretion. If, however, the gland is transplanted under the renal capsule, then hypothalamic control is lost and the secretion of anterior pituitary hormones falls, save that of prolactin which increases.
[A:465; I:738]

398 Growth hormone, prolactin, ACTH and MSH are all single-chain polypeptides. TSH, LH and FSH are all double-chain glycoproteins. Similar hormones are produced by the human placenta. For example, placental lactogen is similar in nature to growth hormone and prolactin, and human chorionic gonadotrophin resembles pituitary LH.
[A:464]

399 The majority of anterior pituitary hormones exert their effects on other endocrine glands. The hormones produced by these glands feed back at the pituitary or hypothalamic level. Growth hormone and prolactin exert metabolic effects on

123

tissues other than endocrine, growth hormone having the more widespread effect. Thus there is no hormonal feedback on their release. Interestingly, these two hormones have release controlled by inhibitory factors, although a growth-hormone-releasing hormone also exists.
[I:740; J:205]

400 Growth hormone has widespread effects, including the increased transport of amino acids into cells and incorporation into proteins. It inhibits fat synthesis and facilitates the release of fatty acids from adipose tissue. In the adult, excess causes growth of tissues such as the heart and connective tissue so that the hands and feet enlarge. Growth hormone stimulates the growth of the long bones before the epiphyses have fused so that excessive growth is seen in the child, whereas in the adult there is, for example, growth of the jaw bone.
[C:314; D:161; E:998]

401 In addition to its other effects, growth hormone influences carbohydrate metabolism, inhibiting uptake of glucose into cells and decreasing the sensitivity to the hypoglycaemic effect of insulin. Thus growth hormone has a diabetogenic action.
[C:321; F:494]

402 In addition to feedback control of anterior pituitary hormones, there is control from higher centres. The hormones are not released continuously, but in spurts. The size and frequency of the spurts of growth hormone increase with the onset of sleep. In man, ACTH secretion reaches a peak in the early morning and decreases to its lowest levels at midnight. Both hormones are additionally released in response to a number of factors, including stress.
[A:468; C:316]

403 Both ACTH and TSH maintain the structure and function of their target glands and increase blood flow to the organs. Excess secretion leads to hypertrophy.
[E:1012; G:196; I:771]

404 The effect of growth hormone on bone and cartilage formation results from the mediation of a group of substances known as the somatomedins. Evidence for the formation of these factors

124

was provided by the observation that growth hormone did not stimulate growth of cartilage *in vitro* whereas serum of intact animals did.
[A:469; H:1475; J:459]

405 Goliath seems to have suffered from loss of peripheral vision, which might be expected if a pituitary tumour were present.
[C:321]

406 With the development of radioimmunoassay, direct determination of plasma concentrations of pituitary hormones may be made. There is considerable variation in the resting hormone concentrations, so challenge tests using, for example, insulin may be employed to investigate growth hormone, prolactin and ACTH, while LHRH and TRH may be used for the gonadotrophins and TSH. Radiography of the skull may also be used to detect the presence of a tumour.
[A:466; C:321]

407 The two hormones released from the posterior pituitary are oxytocin and vasopressin (anti-diuretic hormone). They are nonapeptides, differing in only two amino acids in positions 2 and 8.
[A:471; B:7-26; D:160; E:1003; H:1486]

408 Oxytocin and vasopressin are synthesised in the supraoptic and paraventricular nuclei, in the form of precursor molecules which also comprise the so-called carrier protein, neurophysin.
[B:7-26; E:1000; F:494; H:1485; I:734]

409 A change in the osmotic pressure of the extracellular fluid appears to be the predominant physiological factor in vasopressin release. The receptors involved are the osmoreceptors located in the anterior hypothalamus.
[I:735; J:388]

410 Vasopressin acts on the final part of the distal tubule and on the collecting duct to increase the permeability to water, a process involving cyclic AMP. Water then passes out of the renal tubule into the interstitium where the osmotic pressure is higher. Thus an antidiuresis is promoted.
[A:253; C:171; H:1486]

411 Vasopressin acts to constrict the blood vessels, causing an increase in blood pressure and to slow the heart. This action is of importance in haemorrhage and it has also been suggested that vasopressin plays a part in hypertension.
[C:458; F:495]

412 Oxytocin acts on the uterus, stimulating contractions, and is thus important in the delivery of the fetus. It also acts on the myoepithelial cells and the mammary glands to promote milk ejection.
[C:160; E:1002; G:205; H:1489]

413 No known syndrome is associated with over- or underproduction of oxytocin, but overproduction of vasopressin results in the syndrome of inappropriate production of antidiuretic hormone (SIADH), characterised by low plasma osmolality and concentrated urine. Undersecretion results in diabetes insipidus in which up to 20 litres of dilute urine may be produced per day.
[A:253; B:7-31; D:160; E:1002; H:1489]

414 The intracellular concentration of iodide in the thyroid is some 20–50 times greater than plasma iodide levels. This concentration gradient is maintained by an active pump in the follicular cells.
[B:7-34; E:1006; H:1499; I:768; J:196]

415 After the iodide has been taken up into the cells of the thyroid gland, it is bound and oxidised by the peroxidase enzyme which transfers the activated iodide to a tyrosine residue in the thyroglobulin molecule. This results in the formation of mono-iodotyrosine (MIT) and di-iodotyrosine (DIT). Synthesis of the hormone involves internal coupling of one molecule of MIT and one of DIT to give tri-iodothyronine (T_3) and two DIT residues to give thyroxine (T_4). The thyroglobulin molecule acts as a store for preformed T_3 and T_4.
[B:7-35; E:1007; F:511; H:1499; J:196]

416 TSH stimulates iodide turnover in the folicular cells, but first causes an immediate loss of iodide and then an increase in the active uptake of iodide by stimulating synthesis of the protein involved in the active transport of iodide. The hormone also has a general metabolic effect on

the thyroid, increasing the synthesis of thyro-globulin and thyroid hormones. Finally, TSH increases the breakdown of colloid and release of thyroid hormone from the gland.
[B:7-39; E:1011; G:200; H:1513; I:771]

417 The hormone T₃ would be more effective than T₄ in relieving symptoms.
[A:473; C:248; H:1507]

418 The hormones T₃ and T₄ act at the level of the cell membrane to cause increased protein synthesis and a consequent increase in the levels of enzymes. This leads to an increased rate of oxygen consumption.
[C:248; H:1509; I:775]

419 The normal development of the central nervous system is dependent on a normal thyroid function during pre- and postnatal life. If a child is hypothyroid, there will be failure in both growth and development of the central nervous system, the term 'cretin' being used to describe the condition.
   In the adult patient, hypothyroidism or myxoedema is characterised by reduced metabolism, so the patient exhibits slowness of thought and action and impaired memory. The skin is dry and thick, there is loss of hair and intolerance to cold. The patient may also be constipated.
[C:252; G:201; H:1495]

420 A goitre is an enlarged thyroid gland. Endemic goitre is associated with certain geographical areas and is produced by a prolonged deficiency of iodine in the diet. Goitrogens are substances which produce goitres and may be present in the diet.
[A:475; F:517; G:200]

421 In hyperthyroidism there is an increase in metabolic rate with an increase in body processes in general and weight loss. There is an increase in heart rate and cardiac output, and atrial fibrillation may occur, especially in older patients. Normal mood variations are exaggerated, so patients may be nervous and irritable. Exophthalmos is also noted.
[C:253; F:518]

422 There can be present in the blood a thyroid-stimulating factor which has a much longer half-life in blood than TSH. It was initially termed long-acting thyroid stimulator (LATS). It is a globulin capable of binding to the thyroid gland and stimulating the release of thyroid hormones.
[C:253; D:163; H:1502]

423 Thyroid function can be assessed in two ways. In the first, a radioactive isotope of iodine is injected and its uptake into the thyroid gland is measured using a calibrated detector placed on the neck. The second method is the use of radioimmunoassay to measure the plasma concentrations of TSH and of T_3 and T_4.
[B:7-43; D:164; E:1015]

424 The diagram should show that the gland comprises two parts, which are anatomically and functionally distinct—the central medulla and the outer cortex. The medulla produces catecholamines. The cortex comprises three zones: the outer zone is termed the zona glomerulosa, the next the zona fasiculata and the inner zone the zona reticularis. The most important hormone produced by the inner two zones is cortisol. Aldosterone is the most important hormone produced by the zona glomerulosa.
[A:476; E:1030; G:192; H:1558]

425 It is the adrenal cortex which is essential to life. The function of the adrenal medulla can be compensated for by an increase in sympathetic activity.
[C:277; H:1558; J:192]

426 Control of the secretions of the adrenal medulla is nervous, being controlled from the hypothalamus. Stimuli causing release include emotions, stimulated muscle activity and hypoglycaemia.
[C:292; I:782]

427 The three catecholamines are all derivatives of tyrosine, which is first hydroxylated to L-dopa which is then decarboxylated to dopamine. Hydroxylation of dopamine yields noradrenaline (norepinephrine) which may be methylated to give adrenaline (epinephrine).
[D:165; J:173]

428 Catecholamines help to maintain blood glucose levels, influencing both glycogen and fat metabolism and gluconeogenesis.
[F:395; J:454]

429 Steroid hormones have as their basis three six-carbon rings and one five-carbon ring. Those hormones with 21 carbon atoms (C-21) possess glucocorticoid and mineralocorticoid activity. The C-19 steroids are sex steroids. A ketone group on position 3 and a double bond between positions 4 and 5 are essential for the activity of all groups.
[B:7-55; C:281; H:1559; J:33]

430 Cortisol is the main steroid with glucocorticoid activity, and in man is transported in the blood bound to a globulin called transcortin or corticosterone-binding globulin. When this protein is saturated, cortisol binds to albumin.
[A:478; B:7-56; G:195; H:1565]

431 ACTH can influence aldosterone production by the zona glomerulosa. The most important influence on the secretion of aldosterone is angiotensin II. It has also been observed that potassium ions can directly affect aldosterone production.
[A:479; E:1023; H:1588; J:790]

432 Primary aldosteronism, also called Conn's syndrome, is due to increased secretion of aldosterone with no increase in cortisol secretion. The symptoms are due to excessive amounts of aldosterone, a hormone which acts mainly on electrolyte balance. Thus hypokalaemia is observed (reduced serum K^+), leading to tetany. There is an increase in the serum sodium, which causes a rise in the extracellular fluid volume and, consequently, hypertension. The difference between primary and secondary aldosteronism is that the latter is brought about by extra-adrenal factors such as oedematous conditions occurring in congestive heart failure, nephrosis and toxaemia of pregnancy.
[B:7-67; D:165; F:509]

433 The symptoms of Cushing's syndrome are those associated with oversecretion of cortisol. There is redistribution of the body fat such that the face

becomes moon-shaped and fat is deposited at the back of the neck and over the abdomen. The face becomes red and there is thinning of the bones and skin and reduced collagen in the blood vessels, so red striae occur with easy bruising. There are also changes in carbohydrate metabolism, and in women there may be some degree of virilism and amenorrhoea.
[C:299; E:1033; F:508]

434 The adrenogenital syndrome results from an increase in the level of androgens, which causes masculinisation in the female and precocious puberty in the male. It results from a specific enzyme deficiency, most commonly 21-hydroxylase. The synthesis of steroids at the end of the pathway cannot be completed and intermediates such as 17-hydroxypregnenolone accumulate and spill over into the production of androgens. Furthermore, both dehydroepian-drosterone and androstenedione are converted in the peripheral tissues to testosterone. The lack of feedback from cortisol results in elevated ACTH concentrations.
[B:7-65; F:509; H:1594; I:791]

435 In Addison's disease there is a chronic insufficiency of the adrenal gland with a reduction in the secretion of cortisol and aldosterone. The state is characterised by debility and nausea accompanied by weight loss, pigmentation of the skin and mucous mem-branes. Even though the blood pressure is low there is a high urine output and an increased sodium loss in the urine, so there is volume depletion and a reduced plasma sodium.
[C:300; E:1032; F:507]

436 Plasma calcium concentration is 2.5 $mmol \cdot l^{-1}$. Nearly 50 per cent is present as the free calcium ion, 45 per cent is bound to protein (especially albumin) and the remainder is complexed as the salts of organic acids such as citrate, phosphate and bicarbonate.
[A:283; B:7-47; D:164; H:1520]

437 Calcium is involved in nearly every biological function. The circulating concentration of calcium is particularly important, as it influences the threshold for excitability of cell membranes in nerve and muscle and the amount of

transmitter released from nerve terminals. It also influences the coupling of excitation to processes such as the contraction of muscle and secretion from glands. Amongst other functions, it is important in enzyme reactions, the clotting of blood and bone formation.
[C:301; H:1519; I:758]

438 The majority of calcium in the body is found in the bones—about 1 kg—largely as crystals of hydroxy apatite $(3Ca_3(PO_4)_2Ca(OH)_2)$. Only some 5 g of calcium in the bone can be rapidly metabolised.
[H:1523; I:758; J:393]

439 Both calcium and phosphate are filtered in the kidney tubules and the rest is actively reabsorbed, a process hormonally controlled. Thus parathyroid hormone increases reabsorption of calcium while increasing excretion of phosphate.
[A:284; C:302; E:1052; F:522; H:1523]

440 The plasma calcium concentration is important in the secretion of both parathyroid hormone, which is inversely related to calcium concentrations, and calcitonin, which is directly related to them.
[F:522; J:394]

441 Parathyroid hormone is important for life, removal of the gland leading to hypocalcaemic tetany in which spasm of the larynx is so severe that asphyxia ensues. Calcitonin, on the other hand, is not vital for life.
[D:164; F:523]

442 Parathyroid hormone raises plasma calcium by increasing bone resorption, increasing reabsorption of calcium from kidney tubules and increasing absorption of calcium from the gut. Also, through its action on bones, it increases the concentration of phosphate in the plasma, but in turn increases the amount of phosphate in the urine.
[A:285; B:7-48; H:1541; I:759]

443 Calcitonin acts on the bone, probably reducing the mobilisation of calcium. Excretion of hydroxyproline, which comes from the collagen matrix of bone, is a good indicator of bone

resorption; hence it would fall if calcitonin inhibits bone resorption. Partly as a result of its action on bones, it also lowers circulating calcium. It also lowers phosphate levels.
[D:308; H:1551; I:761]

444 Vitamin D is derived from cholesterol. A metabolite formed in the kidney, 1,25-di-hydroxycholicalciferol, appears to be the active form. Its production in the kidney is partly regulated by parathyroid hormone.
[A:287; B:7-50; C:303; H:1531]

445 Vitamin D is required for normal bone growth and the absorption of calcium in the intestine, so if a deficiency occurs in the growing animal there is inadequate calcification of bone and it tends to become deformed. Permanent deformation of bone is seen in rickets, which may result from vitamin D deficiency.
[F:525; H:1531; J:395]

446 The primary substance regulated by the mechanisms of the postabsorptive state is glucose.
[C:219; J:447]

447 The hormones are insulin, secreted from the β or B cells, and glucagon, from the α or A cells.
[B:7-74; D:168; H:1639; I:745]

448 The insulin molecule consists of two chains, A and B, comprising 21 and 30 amino acids, respectively, and linked by two disulphide bridges. It is synthesised in the form of a precursor molecule consisting of 86 amino acids. Glucagon is a polypeptide of 29 amino acids and, again, may be synthesised in the form of a precursor.
[A:91; B:7-76; G:207; H:1661]

449 Insulin is released in the fed state, when the circulating glucose concentrations are high, and its action is to reduce the glucose concentrations. Glucagon, growth hormone, adrenaline, ACTH and, hence, glucocorticoids are all released in the fasting state, when circulating glucose concentrations are low. All tend to elevate blood glucose.
[A:90; C:268; I:753]

450 Blood glucose depends on a balance between a number of factors, including absorption from the intestine, uptake in peripheral tissues, metabolism in the liver and absorption or excretion by the kidneys.
[G:206; J:442]

451 Insulin has four direct effects on organic metabolism: stimulating uptake of glucose by muscle and adipose tissue; the formation of glycogen from glucose in liver and muscle; the formation of triglyceride from glucose in liver and adipose tissue; and inhibiting release of fatty acid from adipose tissue. It also stimulates the entry of potassium into cells.
[A:90; D:168]

452 In the absence of insulin there is a failure to metabolise ingested glucose, so blood glucose concentrations rise and glucose appears in the urine, inducing a solute diuresis. In the absence of insulin, the reactions of other hormones influencing carbohydrate metabolism operate without restraint. Thus adrenaline and glucocorticoids stimulate the breakdown of glycogen in the liver, and glucocorticoids mobilise muscle amino acids with rapid formation of glucose by the liver; these actions lead to elevated plasma glucose and causes a high rate of urea formation and hence increased nitrogen excretion. Adrenaline, growth hormone and ACTH stimulate fatty acid release from adipose tissue, so there is a massive production of acid ketone bodies. These last are excreted in the urine in association with cations, so there is loss of sodium and, more significantly, potassium.
[C:265; F:490; H:1641; J:451]

453 Juvenile onset diabetes is insulin dependent and may be in part hereditary or associated with infection. Maturity onset diabetes is insulin independent and may be associated with other endocrine disorders.
[A:92; B:7-84; C:275]

454 The patient is fasted before the test and the fasting blood glucose level determined. He is then given a glucose load orally (generally 50 g) and the blood sugar determined at intervals. At about 30 minutes in the healthy individual the glucose rises to a peak, usually not great than 10

133

mmol·l⁻¹ (180 mg·100 ml⁻¹), and should return to normal within 2 hours. In a diabetic subject there is a large and prolonged rise in blood glucose.
[A:92; B:7-85; F:488; H:1658]

The reproductive system

455 Genetic sex is determined by the number of sex chromosomes, and can be established by demonstrating the presence of sex chromatin (Barr body) or by chromosome determination in cultured cells.
[B:7-86; C:322; I:793]

456 The diagram should show the testes, epididymis, vas deferens, the seminal vesicle, prostate gland, bulbourethral (Cowper's) gland and penis.
[A:486; B:7-93; E:1072; J:480]

457 Spermatogenesis occurs in the seminiferous tubules. The innermost layer of cells, the spermatogonia, move away from the basement membrane and increase in size to form the primary spermatocytes containing 46 chromosomes. They divide to form the secondary spermatocytes which have 23 chromosomes. These divide to form the spermatids which ultimately become the mature spermatozoa. FSH acts on spermatogenesis via stimulation of the Sertoli cells, and LH is important as it stimulates testosterone which is required for spermatogenesis.
[A:484; B:7-106; E:1073; H:1624; J:481]

458 Spermatogenesis is thought to be adversely affected by a high temperature and this may be seen if the testicles fail to descend (cryptorchidism).
[C:334; E:1075; G:457; H:1628]

459 Testosterone is the natural anabolic steroid; it promotes the retention of nitrogen and electrolytes, protein synthesis, growth and strengthening of skeletal muscle. It has a widespread effect and elevates the metabolic rate.
[B:7-104; G:456; H:1631; I:799]

460 The morphology and function of the accessory reproductive organs and of spermatogenesis depend on testosterone, as do nearly all the obvious male secondary characteristics such as hair growth, deepening of the voice, skin texture and masculine pattern of muscle and fat distribution.
[A:486; B:7-102; D:171; E:1080; H:1630]

461 The volume per ejaculate is 2.5–3.5 ml and the sperm count $100 \times 10^{-6} \cdot ml^{-1}$, a count lower than $20 \times 10^{-6} \cdot ml^{-1}$ resulting in infertility. The seminal fluid is rich in prostaglandins and in fructose, a source of metabolic energy for the spermatozoa. The prostate gland secretes a thin fluid containing calcium citrate and acid phosphate, while the bulbourethral gland secretes mucus.
[A:485; D:172; E:1076]

462 The diagram should show the ovaries, fimbriae, oviducts, uterus, cervical canal and vagina.
[C:340; E:1086; J:492]

463 Just before ovulation the mature ovum divides. The formation of the ovum is similar to that of the primary spermatocyte in that during a meiotic division each daughter cell receives 23 chromosomes. However, one cell retains virtually all the cytoplasm. The ovum matures in the Graafian follicle under the influence of the gonadotrophins. One follicle grows until it ruptures, extruding the ovum, and the other follicles degenerate. At the time of ovulation the follicle fills with blood and then is converted to the corpus luteum where oestrogen and progesterone are both synthesised.
[D:170; J:495]

464 The first day of the menstrual cycle is conventionally taken as the first day of menstruation, when the endometrium is shed. The regrowth of the endometrium (proliferative phase) starts on day 5, under the influence of oestrogen, and extends until ovulation on approximately day 14. Then coiled blood vessels and secretory glands develop under the influence of progesterone and the secretory phase is entered. Finally, the spiral arteries constrict followed by their rupture, producing haemorrhage and sloughing of tissue.
[A:489; G:439; H:1612]

465 At the beginning of menstruation there is a slight elevation in the circulating concentrations of oestrogens. These hormones exert a negative feedback on the release of gonadotrophins, but a rise in their concentrations which is maintained before ovulation produces a positive feedback on the hypothalamus and there is a surge of LH accompanied by FSH. During the luteal phase, high concentrations of oestrogen and progesterone produced by the corpus luteum have a negative feedback on the gonadotrophins.
[C:345; E:1087; H:1614; I:807]

466 There is an increase in body temperature of 0.2–0.6°C (0.4–1.0°F) at the time of ovulation, probably as a result of increased progesterone concentrations.
[A:492; B:7-113; D:171]

467 As well as development and maintenance of the uterus, oestrogens are required for the development of the cervix, vagina and labia majora and minora. Oestrogens affect the vagina, causing cornification, and act on the cervical mucosa so that a thinner, more alkaline, mucus is produced. Mucus is thinnest at ovulation and a cervical smear will form a fern-like pattern. Oestrogen helps to produce duct proliferation in breast tissue at puberty and the alveolar development during pregnancy.
[C:344; E:1092; G:440; H:1608; J:501]

468 Under the action of progesterone the vaginal walls are thickened and the cervical mucus becomes thick and cellular. In the breast—along with oestrogen and other hormones including cortisol, growth hormone and prolactin—it stimulates development of lobules and alveoli.
[E:1094; G:442; H:1611; I:807]

469 The combination pills contain oestrogen with progesterone. They are thought to act by depressing secretion of gonadotrophin-releasing hormone and hence LH secretion, so that ovulation does not occur. It has been suggested that ovulation may occur in some women, but that the progesterone makes the cervical mucus thick and unfavourable to sperm migration, and may also interfere with implantation.
[A:492; E:1101; G:446]

470 Menopause is the stage when menstruation ceases and is due to failure of ovarian function. There is a fall in the circulating concentrations of oestrogen and progesterone, which causes a number of changes in the body—especially a fall in protein anabolism. As the negative feedback on the pituitary is removed, there is a pronounced release of gonadotrophins.
[A:578; F:537]

471 The time of ovulation varies from individual to individual, but is around day 14 and for 10–15 hours afterwards fertilization may occur. Implantation occurs some 7 days later.
[H:1619; I:812; J:503]

472 In the menstrual cycle, the corpus luteum begins to degenerate about 4 days before the next menses and is replaced by scar tissue to form a corpus albicans. In pregnancy, the corpus luteum enlarges in response to stimulation by production of gonadotrophic hormones secreted by the placenta and itself secretes oestrogen, progesterone and relaxin. The corpus luteum maintains the pregnancy for the first 6 weeks, the progesterone maintaining the endometrium in the required state, and its function begins to decline after the first 8 weeks of pregnancy.
[G:450; I:872]

473 Characteristic of pregnancy is the production by the placenta of human chorionic gonadotrophin which is similar in structure to pituitary gonadotrophin. It can be detected in the blood as early as 6 days and in the urine as early as 14 days after conception, and hence its identification may be used as a test of pregnancy. It peaks at 60–80 days, falling away by 3 months. Human placental lactogen (HPL), oestriol and pregnanediol all increase progressively through pregnancy, peaking just before delivery. If oestrogen and HPL do not increase normally, this may be an index of fetoplacental insufficiency.
[A:497; B:7-118; C:351; E:1109; F:546; H:1619]

474 In addition to being important in the synthesis of a number of hormones, the placenta serves as the embryo's lungs, gastrointestinal tract and kidneys.
[A:499; F:546; G:449]

475 As the placenta serves a number of functions, certain tissues in the fetus may be bypassed. Most of the blood returning to the fetus via the umbilical vein bypasses the liver in the ductus venosus and travels to the right atrium. Blood does not pass through the lungs, but instead reaches the left side of the heart via the foramen ovale between the right and left atrium and ductus arteriosus which transfers blood directly from the pulmonary artery to the aorta. Blood travels to the placenta via the umbilical artery and so bypasses the digestive tract.
[A:503; F:550; H:1966]

476 In parturition the spontaneous contractions of pregnancy become powerful and dilate the cervix. In the first stage of labour, regular contractions start and become more pronounced. The infant is delivered in the second stage, and in the third stage the placenta and membranes are expelled. In some animals the trigger for parturition comes from the fetus, but it is not clear if this is the case in man. An increase in oestrogens and a fall in progesterone are important for enhanced uterine activity, as are prostaglandins and oxytocin. Oxytocin is sometimes used to induce labour.
[E:1115; G:451; H:1622; J:509]

477 Prolactin in the presence of insulin, adrenal corticosteroids and thyroid hormones is responsible for milk secretion. Oxytocin produces contraction of the myoepithelial cells and hence expression of milk from the mammary gland.
[C:352; D:175; F:554; H:1623]

478 The main protein constituent of cow's milk is caseinogen which is not as readily digested as the protein in human milk, lactalbumin. Also, cow's milk has more carbohydrate and less salt.
[C:353; E:1120; F:555]

479 Throughout the body of the newborn is a type of fat known as brown fat, which is highly vascular and contains fat cells rich in mitochondria. Increased metabolism within this tissue causes additional production of heat and thus helps to maintain body temperature.
[A:419; D:175]

The nervous system

Sensory input

480 Information can only be transmitted by action potentials which are virtually identical throughout the central nervous system. The specificity of the receptors, the afferent pathways and the brain areas they supply allows the type of sensation and its location to be identified. Intensity of stimulation is reflected by the frequency of action potentials and the number of receptors responding.
[A:347; B:8-10; E:641; H:336; I:61]

481 Stimuli from the eye are encoded in terms of light. Therefore any stimulus such as a blow results in the impression that light has been seen.
[C:64; H:336; J:558]

482

Sensory receptor	Probable cutaneous sensation
Hair follicle endings	Touch
Nacked nerve endings	Pain, temperature, touch
Expanded ends of sensory fibres:	
Merkel's discs	Touch, pressure
Ruffini's corpuscles	Touch, warmth
Encapsulated endings:	
Krause's end-bulbs	Touch, cold
Meissner's corpuscles	Touch, pressure
Pacinian corpuscles	Touch, pressure, high frequency vibration

[D:191; G:357; H:341]

483 The different types of receptor are mechanoreceptors (e.g. muscle spindles), chemoreceptors (e.g. olfaction), thermal receptors and visual receptors.
[A:352; C:64; E:641; F:281; H:329]

484 When a receptor is activated, the permeability properties of the membrane are increased so that the receptor membrane depolarises (i.e. the membrane potential is reduced), giving a generator potential; the greater the stimulus, the greater the degree of depolarisation. Thus the response is graded and does not obey the all-or-none law.
[B:8-8; G:351; H:330; I:64]

485 In contrast to the action potential, generator potentials can be added, have no refractory period and a duration of greater than 1–2 ms.

Furthermore, they are conducted passively and, unlike action potentials, decrease in magnitude with distance.
[G:357; J:152]

486 Adaptation is a decrease in the frequency of the action potentials in the afferent nerve in the face of continuing stimulation.. Phasic receptors adapt rapidly and, as they respond primarily to a change in stimulus, are responsive to a change in the environment. Tonic receptors adapt slowly and thus give constant information as to the relation of the body to the surroundings.
[A:354; D:192]

487 (a) Dorsal columns carry proprioception and discriminatory sensation. The first order neurones pass to the nucleus gracilis and cuneatus, and those fibres entering the column first (i.e. sacral fibres) lie medially. The second order neurones cross in the lower medulla and ascend to the ventral posterior nucleus of the thalamus, while the third order neurones pass to areas 1, 2 and 3 of the sensory cortex.
(b) The dorsal spinocerebellar tracts carry unconscious proprioception. The first order neurones synapse in Clarke's column in the dorsal grey matter and then pass to the cerebellum.
(c) The lateral spinothalamic tract is the pathway for pain and temperature. Within several segments of entering the cord, the fibres synapse, cross and ascend to the thalamus. Those fibres entering the cord later (i.e. from the cervical region) run medially.
(d) The ventral spinocerebellar tracts which have relatively little representation from the lower parts of the body carry unconscious proprioception. First order neurones synapse at the level of entry, and second order neurones cross to ascend to the cerebellum. Some fibres may cross again to the ipsilateral side.
[B:8-143; C:79; D:212; F:284; H:353]

488 In general terms, the dorsal column system transports information which must travel rapidly and with temporal fidelity. It carries information that is discretely localised and sensations detecting fine gradations of intensity. In contrast, the spinothalamic system transmits a broad spectrum of information, temperature,

pain and some (crude) touch. The dorsal system is confined to proprioception and touch.
[A:363; D:192; E:652]

489 Peripheral nerves contain sensory nerves carrying all types of modalities as well as motor and autonomic fibres. There is thus a variable sensory loss as well as a loss of muscle stretch reflexes for the area supplied by the nerve. At the supratentorial level all major sensory pathways have crossed to the contralateral side, so lesions here result in loss of sensory function on the entire contralateral side.
[A:358; F:284]

490 If the spinothalamic tract is interrupted, the sensations of pain and temperature are lost in all segments below the level of the lesion. This knowledge may be used in cases of intractable pain.
[C:80; G:410]

491 In the Brown-Séquard syndrome there is effectively hemisection of the spinal cord, resulting in the loss of the sensations of pain and temperature on the opposite side to the lesion, and of proprioception and motor function on the same side as the lesion. The lesion is frequently incomplete.
[A:371; D:202; H:384]

492 There is a 'point for point' localisation of perceptive areas in the sensory cortex. Stimulation of various parts of the postcentral gyrus results in sensations arising apparently from the appropriate area of the body. There is a high density of receptors in the face and hands, so these have a large representation in the sensory cortex.
[C:81; E:654; J:563]

The eye

493 The diagram should show the following: the cornea, aqueous and vitreous humours, crystalline lens, ciliary muscle, pupil, iris, canal of Schlemm, optic nerve, fovea and blind spot.
[A:424; B:8-63; H:482; I:106]

494 The highly vascular ciliary processes of the ciliary body of the eye are believed to secrete the

141

aqueous humour. It is normally absorbed via the canal of Schlemm, passing down in fine ducts to the intrascleral venous plexus. If this outlet is obstructed—either by decreased permeability of the trabeculae or by forward movement of the iris—intraocular pressure is increased and the eye disease glaucoma results.
[A:425; C:90; H:483]

495 The lacrimal glands secrete tears, subserving a protective function, preventing dehydration and washing away irritating particles. Tears also contain a bactericidal enzyme, lysosome.
[A:424; I:107]

496 Light is refracted at the cornea and this is where the greatest degree of refraction occurs. Some refraction occurs at the lens and, although it contributes relatively little to the focusing of the image, adjustments for distance are made here. When the lens is removed, the power of the eye, measured in dioptres (l/focal length in metres) is reduced from 59D to 43D.
[B:8-63; D:212; E:788; F:359]

497 In accommodation the lens is thickened to allow near objects to be viewed. This is brought about by the contraction of the ciliary muscle which encircles the lens, which in turn decreases the tension on the suspensory ligaments which are attached to the periphery of the lens. The lens is thus allowed to assume a more spherical shape. There is also constriction of the sphincter-like papillary muscle, which prevents rays passing through the peripheral parts of the lens which might limit clarity of vision.
[D:212; E:788; F:359; H:488]

498 Hyperopia or far-sightedness is generally the result of a short eyeball, but may be due to too small a refractive power of the optical system. It may be corrected by using a spherical convex lens. Myopia or short-sightedness results from an abnormally long eyeball or, less commonly, too great a curvature of the cornea and lens. A concave lens may be used to correct vision. Astigmatism is due to abnormal curvature of the cornea and may be corrected with a cylindrical lens.
[A:427; B:8-65; E:792; H:485; J:568]

499 The near point, the shortest distance at which an object may be seen distinctly, recedes with age until it becomes too far to allow one to read easily. In the young adult the maximum strength of the lens is around 77D. The appropriate lens strength spectacles may be calculated as, when lenses are used together, their combined power may be determined by algebraic summation.
[B:8-64; D:212; H:486; I:116; J:568]

500 When the pupil is exposed to light it constricts, a reflex lost under very deep anaesthesia. Adrenaline causes dilation of pupils (mydriasis) whereas anticholinesterases cause reduction in pupil size (miosis).
[G:386; H:492; I:188]

501 Rods and cones are photoreceptors, so called because of their appearance under the micro-scope. Cones, which are most numerous in the centre of the retina, are needed for colour vision and their visual acuity is high. Rods are more numerous in the periphery of the eye. They detect very small amounts of light, but show only shades of grey. Before reaching the rods and cones, the light must pass the vitreous humour, nerve fibres. ganglion cells and bipolar cells.
[A:430; B:8-69; E:797; F:361; H:504]

502 The Young–Helmholtz theory of vision is based on the three-colour theory. White light and, hence, effectively the sensation of all colours may be produced from lights of three wavelengths: red, green and blue. It is suggested that there are three different cones sensitive to these three colours. Failure of one of these types of cone results in colour blindness in which only two of these three colours are seen.
[G:383; H:514; J:572]

503 The pigment in the rods is rhodopsin, derived from vitamin A, with maximum absorption in the range 505 nm. The rods are responsible for twilight and scotopic vision, so deficiency of vitamin A leads to night blindness. Material from the cones has been identified with an absorption spectra close to 500 nm, and one cone pigment, idiopsin, has been isolated.
[B:8-76; C:97; E:800; I:118]

504 Visual acuity is the smallest distance by which two parallel lines may be separated visually. It is expressed in terms of the reciprocal of the angle the points subtend at the eye. It is tested by using printed letters of such a size that they would subtend an angle of 1 minute if received from a specified distance.
[B:8-99; D:213; G:383; H:489]

505 When illumination is low, a light stimulus causing a subthreshold response in a cone ganglion cell can cause an action potential in a rod ganglion cell so that rod vision is employed. This difference occurs because there is much convergence in the rod pathways, allowing opportunities for spatial and temporal summation. Thus visual acuity is poor and the outline of the object blurred. It is the cones which are concerned with colour vision so that objects seen with rod vision appear in shades of grey.
[A:432; B:8-93; E:805; F:362; H:514; I:121]

506 A focal defect in the field of vision, known as scotoma, may result from lesions in a single optic nerve. The most common lesion of the optic chiasma is in the nasal fibres from the portion of the retina close to the nose so that there is loss of the temporal fields of vision (bilateral hemianopia). Lesions of the optic tract of one side produce homonymous defects in the opposite visual field.
[A:437; I:126]

The ear

507 The diagram should include the tympanic membrane, auditory ossicles (malleus, incus, stapes), round and oval windows, tensor tympani and stapedius muscles, auditory (pharyngotympanic, Eustachian) tube, cochlea and basilar membranes.
[B:8-14; C:106; D:215; E:826; F:351]

508 A pressure level of 0.0002 dyn·cm^{-1} is just at the auditory threshold for the average human and is taken as 0 decibel, normal conversation being 60–70 decibels. Frequencies of 20–20 000 cycles are audible, although the threshold of audibility is not uniform over this range, the ear being most sensitive to sounds of the pitch 1000–4000 Hz.
[B:8-13; C:111; G:366; H:458; J:575]

509 Auditory potentials are carried to the brain via cranial nerve VII, which arises from the spiral ganglion located in the cochlea of the ear, the loudness of the sound being encoded by the frequency of these potentials. The major determinant of pitch is the area of the basilar membrane and hence which receptor hairs are most affected when a sound strikes the ear.
[G:366; I:134]

510 The auditory ossicles transmit the vibrations of the tympani induced by sound, to the oval window of the inner ear. Impaired transmission in the middle ear results in conduction deafness. This can occur in chronic otitis media when the ossicles may be damaged, and in otosclerosis when the attachments of the footplate of the stapes to the oval window become abnormally rigid.
[G:367; I:133]

511 Sensory deafness, in contrast to conduction deafness, is due to disease of the cochlea or the auditory nerve and its nuclei. The two may be distinguished by a number of simple tests performed using a tuning fork—the Weber and Rinne tests.
[A:449; B:8-33; C:115; E:837]

512 The vestibular system is necessary for balance and provides information concerning orientation, rotation and acceleration.
[B:3-38; E:696; F:300; H:813; J:582]

513 The receptors in the vestibular system are hair cells responding to mechanical movement. Within the utricle and saccule are maculae, epithelial cells containing otoliths (calcified particles) which, when the head is turned, distort the hair cells thereby initiating action potentials. The three semicircular canals lie at right angles to each other so acceleration is determined in any plane. Movement also moves the endolymph in the semicirular canals, distorting the cristae and again stimulating the hair cells.
[D:217; H:816; I:143]

514 As the vestibular system is concerned with balance, vertigo—a type of disequilibrium—is associated with disease of this system. Also

experienced are nausea, vomiting, ataxia and nystagmus, the last being a rapid involuntary movement of the eyes. Vestibular nystagmus is asymmetric, with slow movement to the injured side and rapid movement in the other direction.
[A:379; B8-59; G:375]

515 Sound will reach the two ears at slightly different times, which is important in locating a sound of low frequency. The difference in the intensity of the sound reaching the two ears becomes important at higher frequencies.
[A:447; G:367]

Taste and smell

516 Taste may be divided into four sensations—sour, salt, bitter and sweet—each of which has specific locations on the tongue. Sometimes metallic and alkaline are added. Only chemicals which are water- and lipid-soluble affect taste. Substances enter the taste buds via a taste pore and come into contact with the receptor cell membrane.
[A:459; B:8-113; E:839; G:354; H:590]

517 The sensation of taste can be modifed by a number of factors, peripheral and central, many of which are not understood. There is an inherited condition in which people affected cannot taste phenylthiourea. This compound normally has a bitter taste, but 50 per cent of those of Caucasian origin cannot taste it.
[D:218; F:373]

518 Food is identified on the basis of a number of factors, including texture and the odour of the substance. Hence the apparent loss of taste when the nasal passages are blocked in the common cold.
[E:843; J:584]

519 Olfactory receptors are simpler than those associated with taste and, in contrast to the sense of taste, that of smell cannot be divided into different components. Furthermore, a combination of two smells may produce a new one whose constituents cannot be recognised or one smell may obliterate another. The sense of smell can become fatigued. The neural pathways involved in olfaction differ from others in that they have

146

no representation in the cerebral cortex, but pass to the limbic system.
[A:456; B:8-125; J:584]

Pain

520 Pain protects against tissue damage. People without the ability to feel pain suffer from infectious and pathological changes in the spine, bones and joints.
[B:8-146; E:662; G:491]

521 Pain impulses are transmitted to the brain via Aδ fibres 2–5 μm in diameter and C fibres 0.4–1.2 μm in diameter. Impulses in the former are conducted at a rate of 12–30 m·s^{-1} and upon a painful stimulus give rise to a sharp localised sensation. Impulses in the latter are conducted at a rate of 0.5–2 m·s^{-1} and give rise to a dull ache.
[A:350; B:8-148; C:82; E:665]

522 The thalamus is concerned with the affective response to pain, and damage to the nuclei results in the thalamic syndrome in which small stimuli lead to prolonged bouts of pain. In hyperalgesia the sensitivity of the pain receptors is altered, being lower in primary hyperalgesia and elevated in hyperalgesia. Unpleasant sensations are experienced in both conditions.
[C:83; F:381]

523 In contrast to cutaneous pain, visceral pain is poorly localised and associated with autonomic symptoms. It is often referred to other areas.
[C:84; F:375; H:394]

524 (*a*) The specific theory of von Frey assumes that there are specific pain fibres in the tissues.
(*b*) The pattern theory states that powerful stimulation of all receptors results in pain.
(*c*) The gate theory of Melzak and Wall holds that the substantia gelatinosa provides the gate through which pain impulses must pass to reach the lateral spinothalamic system. Inhibition can occur at this site.
[F:379; G:495]

525 Referred pain is felt, not at the site of damage or irritation, but some distance away. Cardiac pain may be referred to the inside of the left arm, and

that from the central portion of the diaphragm to the tip of the shoulder. The pain is usually referred to a region whose embryonic origin was the same as the segment in which the pain arose, the so-called dermatal rule. The convergence theory of referred pain is based on the fact that marked convergence of sensory fibres on the spinothalamic neurones must occur. The facilitation theory postulates that the impulses from visceral structures lower the threshold of the spinothalamic neurones so that the results of minor stimuli are relayed.
[E:668; G:494; I:76]

526 Several neurosurgical procedures may be employed to relieve intractable pain; for example, cutting the connection between the frontal lobes (prefrontal lobotomy). Local anaesthetics block conduction in the periphery by raising the threshold of the neurones. Central anaesthetics are thought to act by interfering with transmission of neural activity in the reticular formation. Drugs such as morphine are analgesics and relieve pain. Morphine receptors are found in the brain which also bind the naturally occurring peptides endorphin and enkephalin.
[G:495; I:68]

The motor system

527 The motor system comprises all the structures controlling the activity of smooth and skeletal muscle. There are four components: the control circuits, the direct (pyramidal) and indirect (extrapyramidal) pathways, and the final common pathway.
[D:194; J:589]

528 The motor unit comprises the α motor neurone, the axon which leaves the spinal cord in the ventral root and the muscle fibres it innervates. The number of muscle fibres innervated ranges from 500–2000 in powerful limb muscles to 5–10 for eye muscles.
[D:185; F:264; H:88]

529 When skeletal muscle contracts in response to stimulation there is reduced tension in the fibres of the muscle spindle, which are in parallel with

the skeletal muscle fibres. The primary receptors in the affected muscle respond to the change in length and rate of stretching, and the secondary endings respond to the change in length, and hence this is reduced. Thus the CNS is informed of conditions within the muscle. [C:44; J:593]

530 A reflex is a stimulus/response sequence which does not entail conscious control. The pathway mediating the response is termed the reflex arc and its components are the receptor, the afferent pathway, the integrating centre, the efferent pathway and the effector.
[B:9-66; D:195; I:147]

531 The diagram should show the intrafusal muscle fibres (nuclear bag fibres, nuclear chain fibres), primary or annulospiral endings, secondary or flower-spray endings, afferent fibres, dynamic γ efferent fibres, static γ efferent fibres, γ efferent fibres, extrafusal muscle and the α motor neurone.
[A:343; B:9-79; E:680; G:362; H:705]

532 The α fibres conduct impulses at 76 m·s^{-1} and the γ motor fibres at 27 m·s^{-1}.
[F:290; G:361]

533 Muscle contraction can be brought about by stimulation of the γ nerve as well as the α nerve. As the intrafusal fibres shorten, the endings are deformed which leads to reflex contraction of the muscle. There is evidence that increased γ discharge may accompany the increased discharge of the α motor neurone so that the spindle can respond to stretch and adjust the rate of firing in the motor neurone during contraction.
[C:74; D:199; J:596]

534 The γ loop is so called because it comprises the γ efferent fibres sending inputs to the muscle spindles and the afferent servofibres back to the central nervous system. The fact that the primary endings respond to phasic as well as tonic changes helps dampen any oscillations which might occur on muscle contraction.
[B:9-79; G:361; J:596]

535 When a muscle is passively stretched, there follows a reflex contraction. The afferent pathway is from the muscle spindle and passes

149

directly to the motor neurone, so this is a monosynaptic reflex. Two examples of a stretch reflex are the knee jerk, in which striking of the patellar tendon produces a reflex contraction of the quadriceps femoris muscle, and the ankle jerk reflex, in which tapping the Achilles tendon produces contraction of the calf muscle and hence plantar flexion of the foot.
[A:367; B:9-70; D:196; H:770]

536 The Golgi tendon organ is in series with the extrafusal muscle fibres and is the receptor involved in the inverse stretch reflex. This is the relaxation of the muscle, seen when the tension in the muscle exceeds a certain level. When muscles are hypertonic, this reflex is particularly marked, being called a clasp knife reflex.
[I:162; J:597]

537 Muscle tone is the resistance of a muscle to stretch, and results from activation of motor units and ensuing stretch of the muscle spindles.
[C:75; G:400]

538 When a muscle contracts, the opposing muscles relax due to reciprocal innervation. The pathway is bisynaptic and involves an inhibitory inter-neurone.
[A:368; E:689; F:271; H:771]

539 The flexor and the crossed extensor reflex are both polysynaptic reflexes. The flexor reflex is a general term describing contraction of the flexor muscle and inhibition of the extensor muscles. It can describe withdrawal of part of the body from a noxious stimulus or flexion during walking. The crossed extensor reflex may accompany the flexor or withdrawal reflex. This last reflex takes precedence over any other.
[B:9-71; C:77; E:687; G:397; H:778]

540 When the sole of the foot of a patient with disease of the motor pathway is stroked with a blunt point, the big toe turns upwards in an extensor plantar response. This sign is called the Babinski response and is the most primitive form of the reflex, being seen in young infants.
[A:387; D:202]

541 Polysynaptic reflexes show (a) irradiation, i.e. the stronger the stimulus, the greater the number of

muscle fibres involved; (*b*) after-discharge, i.e. the stronger the stimulus, the more prolonged the reflex as a result of repeated firing of the motor neurones; (*c*) facilitation, i.e. the reaction time is shortened as a result of spatial and temporal facilitation.
[F:272; I:150]

542 The reaction time is the time between sensory influence and the reflex response, and for the knee jerk reflex in man is 19–24 ms. The central delay is the time for the impulse to go through the spinal cord and is 0.6–0.9 ms. The minimal synaptic delay is 0.5 ms.
[A:364; C:72]

543 Disease of the final common pathway may involve the anterior horn cell, the axon or the motor fibre. No voluntary contraction can be obtained and there is weakness or paralysis with accompanying atrophy. There is also loss of reflexes and tone, a state known as flaccidity. Also associated with disease of the motor unit may be excessive activity leading to fasciculations, cramps and excessive contraction.
[C:75; D:201]

544 Section of the spinal cord is accompanied by spinal shock, a condition in which all cord functions become depressed. There is paralysis, loss of tone and of reflexes, and loss of sympathetic activity, so with sections above L2 there is a fall in blood pressure.
[G:396; I:149]

545 In recovery from spinal shock the neurones recover their excitability, a process which in man may take up to several months. The flexion reflexes reappear first. After several months, hyperexcitability occurs, as may return of muscle tone and tendon reflexes. Patients with major injury of the spinal cord may be left with gross deficits which depend on the level of the lesion. Thus the deficits may vary from loss of the bladder reflex at leves S2–3 to tetraplegia and impaired respiration at levels C4–5.
[A:371; C:140; E:692; H:783]

546 Alternative terms for the corticospinal tract are the pyramidal tract and the direct activation pathway. The term 'pyramidal tract' arose

because the tract forms the medullary pyramids, and the 'direct activation' because the fibres run without synapses from the cerebral cortex to the spinal cord. The main function is to bring about voluntary activity, especially skilled movements.
[D:199; F:310]

547 Specific lesions of the corticospinal tract rarely occur, but such lesions would result in loss of voluntary movements. In contrast to the lower motor lesion, the paralysis is not accompanied by atrophy and the reflexes are maintained. The pattern of loss of function depends of the site of the lesion. Because the fibres cross in the medulla, lesions above this point affect the contralateral side. The motor cortex is immediately in front of the central sulcus of Rolando, and experimental lesion studies show that the body is represented upside down.
[A:370; C:135; H:671]

548 Ataxia is a loss of muscular co-ordination and a severe disturbance of gait. It results from sensory lesions, sensory input being vital to the production of smooth muscular activity. Apraxia is the highest level lesion of motor function in which the ability to perform skilled motor actions by will is lost even though they may still be elicited automatically.
[C:151, 203; E:724; I:237]

549 Muscle tone is important because the degree and distribution determine body posture. It is maintained by the activity of the anterior horn cells.
[F:289; G:422]

550 The main source of sensory afferent impulses for maintaining posture and balance is the eye. There are also contributions from the muscles, joints, skin and vestibular system.
[C:139; J:604]

551 The major co-ordinating systems in the maintenance of balance and posture are the basal ganglia, brain stem nuclei and reticular formation.
[C:139; D:199]

552 In Parkinson's disease there is a failure of the substantia nigra to provide a degree of inhibition of activity, so there are abnormal patterns of rhythmic activity. The disease is thus characterised by a combination of rigidity, akinesia (slow movement) and tremor.
[A:388; F:319]

553 The four necessary components of locomotion are antigravity support for the body, stepping, control of the centre of gravity to allow equilibrium and a means of producing forward motion.
[G:405; I:148]

554 The basic mechanisms for standing and walking lie in the spinal cord. Stimulation of one limb will produce crossed extension of the other limb, followed by extension of the limb originally reflexly flexed.
[A:368; E:690; F:288]

555 The cerebellum is concerned with the co-ordination and smoothing out of movement. It receives inputs from the cortex as to instructions sent to the muscles and information from the muscles as to the response. The cerebellum does not initiate movement nor is it the site of sensory perception. Therefore stimulation does not lead to sensation or movement.
[C:146; J:602]

556 The diagram should indicate that the cerebellum receives information concerning motor commands leading to the spinal cord via the pyramidal and extrapyramidal systems. It also receives sensory information from the muscles and a number of other sources, including the vestibular nuclei. The diagram should also show that corrective signals are sent back to the two motor systems so that output is modified.
[D:199; G:416]

557 The vestibular nuclei are associated with balance, so a lesion in this region would result in difficulty in maintaining an upright posture and associated disturbances of gait.
[A:391; J:602]

558 Decerebrate rigidity is seen in an experimental animal when a section of the brain stem is made

at the level of the colliculi. This results in exaggerated hypertonus and increased reflex excitability. It can be reduced by cooling the surface of the anterior lobe. However, decerebrate rigidity is maintained when the anterior lobe is removed because of excessive α discharge resulting from uninhibited vestibulospinal activity.
[A:374; D:200; E:708]

The autonomic nervous system

559 An alternative name for the autonomic nervous system is the visceral afferent system (also vegetative or involuntary). It is generally divided into the sympathetic and parasympathetic systems, a classification based on the anatomical arrangement of the neurones. Both are two-neurone systems, the fibre of the first neurone being the preganglionic fibre and that of the second the postganglionic fibre. In the sympathetic nervous system the intermediate synapse is situated in a ganglion in the sympathetic trunk which runs either side of the spinal cord so that the postganglionic is a long fibre. In the parasympathetic the synapse is generally situated in the organ supplied. The systems may also be classified according to the transmitter released by the postganglionic fibre—i.e. adrenergic (α or β adrenergic) or cholinergic (nicotinic or muscarinic). In general, noradrenaline is released from sympathetic and acetylcholine from parasympathetic nerves, but some sympathetic fibres such as those to the sweat glands are cholinergic.
[C:153; G:172]

560 The autonomic nervous system supplies smooth and cardiac muscle and exocrine glands.
[A:399; H:894; I:188]

561 The nervous system does not play as important a role in the control of the visceral system as it does in the motor system, because the smooth muscle, cardiac muscle and secretory glands supplied function without external stimulus. There are also systems of local reflexes not involving the spinal cord. In addition, many responses are under the control of hormones, although their secretion is under neural control.

154

Stimulation of the somatic nervous system always leads to excitation of a muscle, whereas stimulation of the autonomic nervous system can lead to either excitation or inhibition. There is also the difference in the anatomical arrangement and the nature of the neurotransmitter.
[H:894; J:185]

562 The preganglionic fibres of the parasympathetic nervous system leave the central nervous system in cranial nerves III, VIII, IX and X, and in the sacral nerves arising from the second, third and fourth segments. The sympathetic fibres leave the thoracic cord over T1–12 and the lumbar cord at L1 and 2.
[A:399; D:205]

563 The autonomic nervous supply to an organ can be purely or mainly sympathetic, mainly parasympathetic or there may be dual innervation. Whatever one division does to the effector organ, the other frequently does the opposite. Innervation by the parasympathetic is relatively discrete as compared with the sympathetic, so the effects of sympathetic stimulation are more diffuse.
[I:188; J:188]

564 The oculomotor nerve (III) supplies the eye, producing accommodation. The facial nerve (VII) supplies the lacrimal gland and the submaxillary and submandibular glands via a complex route, producing enhanced secretion. This is also produced by the glossopharyngeal nerve which supplies the parotid gland. The vagus has the widest distribution, supplying the heart (slowing and producing inhibition of coronary blood flow), the bronchi (constriction and secretion), the stomach and intestines (enhanced peristalsis and secretion) and the proximal colon (vasodilatation and secretion). Thus parasympathetic activity may be said to control functions that are restorative (e.g. digestion).
[C:154; D:206; H:899]

565 Stimulation of the sympathetic nervous system produces vasoconstriction of the arterioles of the body, except in the skeletal muscle where dilatation may occur. It causes sweating and

piloerection in the skin, and through the gastrointestinal tract leads to inhibition of both peristalsis and secretion. However, contraction is produced in the pregnant uterus and relaxation of the urinary bladder. Sympathetic stimulation also increases both heart rate and contractility. The sympathetic system may be said to prepare the body for interaction with the external environment.
[C:159; H:897; J:187]

566 The reflex arc for micturition may be blocked by damage to the sensory fibres, motor fibres or the reflex centre in the conus. This produces a non-reflex bladder. However, damage may occur to the descending pathways or higher centres, leaving the reflexes intact and producing a reflex neurogenic bladder.
[F:385; G:177; H:912]

567 If postganglionic lesions occur, then the axons are likely to degenerate and denervation sensitivity may be exhibited by the eye—i.e. the pupil becomes sensitive to low concentrations of drugs with actions similar to those of acetylcholine or noradrenaline. The postganglionic fibres remain intact with preganglionic lesions and so the paralysed muscle fibres can be stimulated by drugs which displace the neurotransmitter from the nerve terminal.
[A:402; C:161]

568 Ganglion-blocking drugs interfere with transmission at the autonomic ganglia, hence diminishing the tone of the arterioles and reducing peripheral resistance. Adoption of the upright posture produces a decrease in the effective circulating blood volume and the reflex responses are also impaired, so hypotension may be seen.
[C:485; G:243]

569 Atropine blocks cholinergic activity at muscarinic receptors, so one would observe in the patient dilated pupils, a dry mouth, rapid pulse and respiration, a hot dry skin and urinary retention.
[F:389; G:183]

570 The main types of visceral receptor which have been identified are mechanoreceptors and

chemoreceptors. The sources and mechanism of pain in the viscera are not well understood.
[C:79; D:207; H:915]

571 Sensations can reach the central nervous system directly via the cranial nerves or indirectly entering the spinal cord via the splanchnic and pelvic nerves. They are then relayed to the brain stem.
[A:355; D:207]

572 The diagram should show the visceral sensory nerve, the sympathetic ganglion, the somatic sensory nerve fibre and the dorsal root ganglion.
[G:176; H:895; I:149]

573 Control of the visceral system occurs centrally at three levels: the cortex, the hypothalamus and the reticular formation. The hypothalamus provides the major control of reflexes. The part of cerebral cortex and underlying structures which are active in visceral and emotional actions are known collectively as the limbic system (see also question 581).
[A:403; C:140; H:917]

The reticular activating system and electrical activity of the brain

574 In addition to the direct major afferent pathways there is another ascending system—the reticular activating system—which transmits information to the cortex via indirect multineurone pathways. It is concerned mainly with the state of consciousness and also modulation of sensation, with learning and the vegetative functions. It comprises parts of the reticular formation of the brain stem, the ascending projectional system and non-specific projections from the thalamic nuclei to the cortex.
[D:194; J:610]

575 The afferent pathways comprise:
(*a*) branches from the spinothalamic and lemniscal pathways;
(*b*) visceral afferents from the cranial nerves and spinal cord;
(*c*) fibres from the spinal cord;
(*d*) fibres from a number of structures, including the cranial nerve nuclei, basal ganglia and the cerebellum.

The efferent connections include:
(a) the ascending projectional fibres;
(b) the descending reticulospinal fibres;
(c) pathways to the visceral system.
[A:374; C:124; F:298]

576 Destruction of the reticular system results in deep sleep, insensibility to sensory stimuli and immobility.
[F:299; I:170]

577 An EEG is a recording of cortical activity and represents the electrical activity arising from the numerous postsynaptic potentials occurring near the surface in response to cerebral activity as modified by input from subcortical structures. It can be recorded by electrodes placed on the scalp—monopolar records being obtained using a cortical electrode and an indifferent electrode, and bipolar recordings using two cortical electrodes. In the adult at rest with his eyes closed, α rhythms are recorded (with a frequency of 8–13 Hz). Other rhythms are β ($>$ 13 Hz), θ (4–7 Hz) and δ ($<$ 4 Hz).
[A:394; E:734; G:425; H:284]

578 An evoked potential may be recorded following the stimulation of a sense organ, and this may be used to map the specific sensory cortical areas. In unanaesthetised animals or man, the evoked potential is largely obscured by spontaneous activity, but may be recorded by filtering the signal so that the basic rhythm is removed.
[C:125; I:174]

579 Records of EEG may be of value clinically. The two main abnormalities are slow wave or elliptiform. Large waves occur over the brain during an epileptic fit, and abnormal patterns are frequently seen in these patients at other times. Abnormalities may also occur around tumours and hence aid in their localisation.
[A:95; F:323]

580 Sleep may be divided into rapid eye movement (REM) and paradoxical sleep and non-rapid eye movement or slow wave sleep. The EEG recorded during REM sleep resembles that of the wakeful subject. There is also reduced muscle tone and dreaming.
[E:738; G:425; H:307; J:608]

581 The limbic system comprises a ring of cortical tissue on each side of the midline, encircling a number of cortical structures which have a complex series of interactions. These structures include parts of the hypothalamus and thalamus, the septal nuclei, amygdala and hippocampus.
[C:183; E:758; I:228]

582 The limbic system is involved in analysis of olfactory signals, in feeding behaviour and control of several biological rhythms. It is also involved in sexual behaviour and emotional activity, fear and rage.
[C:184; E:761; I:227]

583 Stimulation of electrodes implanted in the limbic system of experimental animals can result in either a pleasant or an unpleasant sensation.
[G:176; J:616]

584 Limbic system disease may manifest itself by effects on olfaction, memory and behaviour.
[I:229; J:616]

585 Learning may be considered to be the ability of the nervous system to store memories or to be a change in response to a given stimulus in the light of experience. Memory is the capability of recalling a thought.
[I:230; J:626]

586 Unconditioned and conditioned reflexes may be illustrated by salivary secretion. The presence of food in the mouth causes production of saliva which is an inborn or natural reflex. If, however, a bell is rung before a dog is presented with food, then eventually the sound of this bell will lead to salivary secretion, which is a conditioned reflex.
[A:392; G:424]

587 The association area occupying part of the frontal lobe is responsible for language expression, that occupying part of the temporal lobe is associated with comprehension of speech, while that occupying both part of the occipital and parietal lobes is associated with the comprehension of written language.
[C:202; J:586]

588 The dominant hemisphere is that cerebral hemisphere in which the interpretive function of the temporal lobe and angular gyrus are most highly developed. These functions are generally only developed in one lobe.
[A:452; E:747; G:424]

589 Memory has two phases: short-term memory, in which it is postulated that the information is stored as reverberating electrical activity in the brain; and long-term memory, which is thought to involve protein synthesis.
[C:201; E:750; J:626]

590 In retrograde amnesia the memory of recent events is lost while earlier memories are unaffected. It can be induced by electroconvulsive shock, low temperature, coma and deep anaesthesia.
[G:423; I:232]

591 Damage to the anterior pole of the frontal lobes causes some change in character and impairment of memory, but the most pronounced change is loss of anxiety. For this reason, prefrontal leucotomy has been performed in patients suffering from an acute state of anxiety. This operation entails severance of connections between the anterior pole of the frontal lobes and the rest of the brain.
[F:331]

CO-ORDINATED FUNCTIONS OF THE BODY SYSTEMS

Acid–base balance

592 Rather than expressing the acidity or alkalinity of a solution directly as the hydrogen ion concentration, it is customary to express it as the pH, which is the negative logarithm to the base 10 of the hydrogen ion concentration.
[D:89; F:7]

593 The pH of a solution with a hydrogen ion concentration of 100×10^{-9} or 10^{-7} mol is 7. If the concentration is 50×10^{-9} mol, it is 7.3 (since the log of 5 is approximately 0.7). Thus pH increases as the hydrogen ion concentration falls.
[D:90; F:7]

594 The range of pH compatible with life is from 7.0 to 7.8, with a corresponding hydrogen ion concentration of 100–16 nmol·l^{-1}. In normal health, the changes observed represent 0.08 of a pH unit.
[C:566; D:90; H:1197]

595 The first defence against shifts of blood pH comprises the buffers in the blood, which act immediately. Second is respiratory compensation, acting in periods of minutes, and finally renal compensation, acting over a period of hours.
[A:268; C:567; H:1735,1794]

596 A buffer system in this context is one which gives up or accepts hydrogen ions. There are three main buffer systems in blood: the bicarbonate system, proteins and haemoglobin. Bicarbonate ions may combine with a hydrogen ion to give carbonic acid or carbon dioxide and water:

$$H^+ + HCO_3^- \rightleftharpoons CO_2 + H_2O$$

Proteins bearing carboxyl and amino groups can give up or accept hydrogen ions. Haemoglobin is an important buffer, as it contains 38 histidine residues capable of donating or accepting hydrogen ions. The hydrogen phosphate/dihydrogen phosphate is also a buffer system, contributing to the response

$$HPO_4^{2+} + H^+ \rightleftharpoons H_2PO_4^-$$

[I:561; J:396]

597 The pK (the pH at which a buffer system is most efficient) for the bicarbonate system is 6.8, which is far removed from the physiological pH. The bicarbonate system is an important physiological buffer, as the carbon dioxide can be blown off in the lungs. At normal pH the ratio of bicarbonate to carbonic acid is 20, but can be altered to maintain the pH, according to the relationship

$$pH = 6.1 + \log_{10} \frac{[HCO_3^-]}{[H_2CO_3]}$$

The term H_2CO_3 may be conveniently expressed by the amount of carbon dioxide in solution.
[G:156; H:1737; J:398]

598 The urinary pH can range between 4.5 and 8.5 which, since this is a log function, represents a

161

vast difference in the concentration of hydrogen ions excreted. Although the respiratory system can help in maintaining the pH constant, it can only excrete carbon dioxide, not hydrogen ions which must leave via the kidney.
[A:268; D:138]

599   In the renal tubular cell, carbonic acid is formed under the influence of carbonic anhydrase and this in turn forms bicarbonate ions which are reabsorbed and hydrogen ions which are excreted as acid and alkaline phosphate. If the ratio of acid to alkaline phosphate is 200 : 1, then the urine pH is 4.5. In addition, ammonia may be produced in the kidney from glutamine and will take up a hydrogen ion to give the ammonium ion.
[F:220; H:1198; J:399]

600   Just as with the excretion of an acid urine the preliminary step is the formation of carbonic acid and hence H^+ and HCO_3^-, so it is with the excretion of an alkaline urine. On this occasion, however, the bicarbonate is excreted and the hydrogen ion retained.
[C:556; I:571]

601   As hydrogen is pumped out of the cell, electrical neutrality is maintained by the inward diffusion of sodium. The sodium is then pumped out by a Na^+–K^+ exchange pump.
[D:138; G:159]

602   If the pH of the blood falls significantly below 7.4, then acidosis is said to be present; if it rises significantly above, alkalosis.
[F:177; H:1201; J:400]

603   If the shift in pH occurs as a result of disturbances in the pattern of respiration, then the condition is defined as respiratory acidosis or alkalosis. If it arises as a result of hydrogen ions being added directly to or removed from the blood, then the condition is defined as metabolic acidosis or alkalosis.
[A:269; D:102; H:1201]

604   If carbonic anhydrase activity is inhibited, then acid secretion by the kidney is inhibited and the actions which depend on it.
[C:560; F:221]

605 If a patient is suffering from a disturbance of acid–base balance resulting from impaired patterns of respiration then the P_{CO_2} will be appropriately altered: elevated to produce respiratory acidosis and depressed to give alkalosis.
[A:268; G:159]

606 In diagnosing an acid–base disturbance it is necessary to know also the base excess, which is positive in alkalosis and negative in acidosis. It is a measure of the acidosis or alkalosis other than that produced by changes in the alveolar P_{CO_2} and is the amount of acid or base which would restore the acid–base composition of 1 litre of blood to normal at a P_{CO_2} of 40 mmHg (5.3 kPa). It can be read off the Siggaard-Andersen nomogram and should not be calculated by estimating the difference between the normal standard bicarbonate (24 mmol·l^{-1}) and the recorded standard bicarbonate. (A reminder—the standard bicarbonate is the value that the bicarbonate concentration would be after elimination of any respiratory component, and not the actual bicarbonate.)
[C:569; I:567]

607 Respiratory acidosis occurs in situations where there is bicarbonate retention such as in bronchitis, emphysema and asthma, and possibly when respiration is depressed as with morphine or barbiturate poisoning. Respiratory alkalosis, on the other hand, occurs in hyperventilation as seen in hysterical overbreathing, at altitude and with salicylate poisoning.
  Metabolic acidosis can occur with ingestion of acids or substances which produce acids such as ammonium chloride and aspirin. It is also seen with increased production of acid as with uncontrolled diabetes mellitus and in exercise. Metabolic alkalosis occurs with acid loss as in prolonged vomiting.
[A:269; G:159]

608 Patient (*a*) with a lowered pH and increased P_{CO_2} is suffering from respiratory acidosis. Patient (*b*) with, additionally, a negative base excess is suffering from both respiratory and metabolic acidosis.
[C:568; I:568]

Haemorrhage

609 The circulating blood volume is about 5 litres, and a haemorrhage of up to 1 litre can be compensated for depending on the rate. With a larger haemorrhage, a blood transfusion is indicated.
[A:104; C:482; H:1082]

610 Haemorrhage produces a reduced stroke volume and hence a fall in blood pressure. This in turn leads to stimulation of baroreceptors and atrial receptors. The chemoreceptors are also stimulated. These events result in increased activity in the vasomotor and cardiac centres, with a consequent fall in the capacity of the circulation and an increase in the heart rate and cardiac contractility. The release of catecholamines augments these reflex changes, and release of vasopressin enhances vasoconstriction. The tissues most affected are those such as skin, kidney and gut.
[D:42; H:1081; I:420; J:309]

611 The increased sympathetic efferent activity to the skin causes marked pallor and inappropriate sweating, so the patient feels cold and clammy. The reduction of renal blood flow leads to marked oliguria.
[B:3-246; C:488]

612 Following haemorrhage, fluid moves into the cardiovascular system from the interstitium. The fall in blood pressure results in a fall in the capillary hydrostatic pressure (and hence the filtering pressure is reduced) and in the glomerular filtration rate. The decreased blood volume results in the elevation of vasopressin levels and hence fluid retention. There is also increased renin release from the renal juxta-glomerular apparatus and of the output of aldosterone from the adrenal cortex, leading to a positive sodium balance and hence increased extracellular fluid volume. Thirst is also experienced and hence there is increased fluid intake.
[I:420; J:321]

613 After the fluid has been replaced, the dilution of the plasma proteins with interstitial fluid and sodium leads to increased albumin synthesis by the liver. The reduction of blood flow to the

kidney initiates the release of erythropoietin, which increases the rate of maturation of red blood cells in the bone marrow and their release.
[A:104; G:478; H:258]

614 Infusion of saline is of limited value, as it distributes through the entire extracellular space. Plasma expanders, which have a high molecular weight and are confined to the circulation, have the disadvantage that they produce additional tissue dehydration. Compatible blood transfusion is the treatment of choice in severe blood loss. Arguments have been put forward for the use of peripheral vasoconstrictors, which decrease the percentage of the cardiac output going to the skin etc., while vasodilators have also been advocated because they alleviate the tissue ischaemia produced by the sympathetic activity and, by lowering the blood pressure, may improve the cardiac output.
[G:480; I:431]

Responses to exercise

615 The efficiency with which a human being works is dependent on a number of factors, including speed, load, fatigue training, etc. However, it is possible to achieve an efficiency of 20–30 per cent, similar to that of an internal combustion engine, but greater than that of a steam engine (10 per cent).
[A:16; G:68; I:704]

616 The oxygen consumption at the beginning of exercise does not increase sufficiently to allow the energy requirements to be supplied by oxidation. Similarly on stopping work, the oxygen consumption does not fall to resting levels. This excess oxygen consumed is termed the oxygen debt.
[G:69; H:1389; I:706]

617 In exercise there is a decrease in the PO_2 of the venous blood and in the oxygen associated with myoglobin and these must be restored. The high energy bonds in phosphocreatine and ATP are rapidly resynthesised and the lactate which accumulated as a result of anaerobic metabolism has to be removed.
[E:947; J:208]

618 Regulation of respiration in exercise does not appear to depend on a single stimulus, but on the combined effects of a number of factors. Apart from the chemical stimuli acting via the arterial blood, it is believed that there are inputs from the cerebral cortex, sensory receptors in the joints and from the rise in body temperature.
[E:567; H:1401; J:359]

619 A number of vasodilator agents act on the muscle during exercise (see the answer to question 159), although a major one is thought to be adenosine. There are also sympathetic vasodilator fibres.
[A:341; J:314]

620 There is a massive sympathetic discharge leading to stimulation of the cardiovascular system, and blood is directed from the kidneys and gastrointestinal tract, etc.—an increase in cardiac output and a rise (although not necessarily) of blood pressure.
[B:3-186; E:371; H:1404; J:314]

621 The most obvious (though not necessarily the most important) contribution to increased cardiac output in exercise is the increased heart rate. Cardiac output also depends on venous return and this is increased by the massaging action on the veins of the exercising muscle and by the mechanical pumping action of the increased respiration. The contribution of the Frank–Starling effect (see question 127) to the response is not clear. Stroke volume may also increase by more complete emptying.
[B:3-121; E:372; G:222]

Responses to high altitude

622 At altitude one is exposed to reduced pressures, reduced oxygen tensions, lowered temperature and extreme variations of weather.
[C:528; E:586]

623 The increased pulmonary ventilation results in a reduced P_{CO_2} and an increase in the pH of the body fluids. These changes inhibit the respiratory centre and Cheyne–Stokes breathing may result (rhythmic waxing and waning of respiration). During the following 3–5 days this

166

inhibition fades, and the dominant effect on respiration is the chemoreceptor stimulus resulting from hypoxia.
[A:235; E:589; H:1859]

624 Hypoxia stimulates erythropoietin, so there is an increase in the packed cell volume from a value of 40–45 to 60–65. Haemoglobin concentrations also increase from a mean of 15 g·dl^{-1} to a value in the region of 22 g·dl^{-1}.
[B:4-23; C:530; G:303]

625 The work capacity is considerably reduced on ascent to high altitude. The capacity can be improved with acclimatisation, but cannot compare with that of subjects normally living at altitude and then taken to a higher altitude. The improvement appears to depend on a high P_{O_2} gradient across the lungs, a high diffusing capacity, improved cardiac responses, a high ventilatory capacity and an increased oxygen-carrying capacity of blood.
[E:590; H:1860; I:479]

626 There is an increase in cardiac activity at altitude and an adaptation of the brain circulation to allow greater blood flow through the hypoxic brain. There is also an increased blood flow to tissues such as the heart and skeletal muscle, and reduced flow to the skin and kidneys. Long term, there is an increased circulating blood volume and vascularity of tissues although the increase in cardiac output is not maintained.
[B:3-119; E:589; G:304]

627 There is a danger of dehydration at altitude, as there is increased water loss via the lungs. However, oedema is also experienced in acute mountain sickness.
[A:235; G:303]

628 The immediate effect of weightlessness on the circulation in man is much like that of assuming a supine posture and there is a moderate reduction in the blood volume. There is also reduced red cell mass, the muscles atrophy and there is reduced work capacity and bone demineralisation.
[E:596]

629 The temperature of the superficial shell of the body varies considerably. The temperature of different tissues within the body also varies, but relatively little so the concept of a relatively stable core or deep body temperature has been adopted. This temperature shows a diurnal rhythm, being lowest at around 3.30 a.m. and highest around 7.30 p.m. The relatively constant temperature is maintained by a balance between heat production and heat loss.
[A:408; E:955; H:1418]

630 The mechanisms for loss and gain of heat are radiation, convection, conduction, evaporation and metabolism.
[B:9-138; E:956; F:334; H:1423]

631 Thermoreceptors are located centrally in the hypothalamus and spinal cord and peripherally in the skin and upper part of the gastrointestinal tract. The main centre for thermoregulation lies in the hypothalamus. In fever the 'thermostat' still works but is set at a higher level.
[A:145; B:9-140; E:961; H:1438; J:378]

632 Heat loss depends on the temperature gradient between the body surface and the ambient air, the prevailing humidity and the wind velocity. The area of exposed body surface is, of course, also important.
[C:180; E:957; H:1425]

633 Changes in skin blood flow markedly influence heat loss. There exist in the skin arteriovenous anastomoses which allow certain capillary beds to be bypassed for a time. Control is via the nervous system and circulating catecholamines. Heat loss from limbs can be minimised by the counter-current heat exchange which allows heat to pass from the arterial blood to the venous blood.
[A:413; H:1424; I:695; J:477]

634 Thermoregulatory sweating occurs at the expense of salt and water balance. Of particular significance may be the loss of sodium.
[E:959; F:336]

635 On exposure to heat, mild hypothermia may result. Large amounts of salt and water are lost,

so the blood volume is reduced and circulatory failure may occur. The lowered plasma sodium also results in muscle cramps. Additional salt must therefore be taken as well as fluid. In a humid climate the effectiveness of heat loss through sweating is reduced. At extreme temperatures heat stroke may result in disruption of brain function and death. Exposure to heat leads to increased cardiac output and cutaneous circulation, frequently with increased venous pressure. It seems that the cause of fatal hyperpyrexia may be the cessation of sweating and circulatory failure.
[B:9-137; E:967; F:336; H:1453]

636 Physiological acclimatisation to heat provides man with thermoregulatory capacities which can be life-saving. There is an altered pattern of sweating and vasodilatation as well as changes in the regulation of water and electrolyte balance and modified work performance.
[F:338; H:1452; I:696]

637 In a cold environment there is reduced heat loss with vasoconstriction and piloerection, which is only effective in furry animals. Instead, man uses protective clothing. Heat may be generated by increased metabolism is response to increased circulating concentrations of thyroid hormones and catecholamines, shivering and exercise.
[F:338; H:1452; I:697]

638 Hypothermia may occur on immersion in cold water or as a result of a cold wind blowing through damp clothing. It may occur in the elderly, as their temperature-regulating mechanism is often defective.
  Hypothermia can also be induced artificially and is of value in certain types of surgery. In such circumstances the circulation may be interrupted for 10–15 minutes to allow surgery on the heart and large vessels.
[C:181; E:968]

Responses to high pressure environments

639 Nitrogen at high pressures produces narcosis, dissolving in the membranes and other lipid structures of the neurones and thus reducing

their excitability. Nitrogen is therefore generally replaced by helium in this situation as it has a much smaller narcotic effect. Oxygen at high pressures can also be detrimental to the central nervous system and may even cause convulsions. Carbon dioxide at high pressures can depress respiration and cause acidosis, but with well designed diving equipment this should present no problem.
[C:555; E:598]

640 After breathing nitrogen at high pressure for a period, large quantities of nitrogen dissolve in the body fluids and tissues. It is not removed, as nitrogen is not metabolised by the body. If at this point a diver returns to the surface, the nitrogen will come out of solution and the bubbles will cause considerable damage (decompression sickness).
[E:600; G:306; H:1929]

641 The phenomenon of supersaturation allows that nitrogen remains dissolved if the pressure of the body fluids is not more than three times greater than on the outside of the body. This means that a diver can ascend from 20.1 m (66 ft, equivalent to a pressure of 3 atmospheres) with no harmful effects. Thus a diver can be brought to the surface slowly in a number of stages.
[A:239; H:1935]

642 On a rapid descent, the volumes of all the gases in the body are greatly reduced, and unless adequate quantities of gas are supplied to the gas-containing cavities—particularly the lungs—serious physical damage results.
[E:604]